THE INFRASTRUCTURAL SOUTH

Infrastructures Series
Edited by Paul N. Edwards and Janet Vertesi

A list of books in the series appears at the back of the book.

THE INFRASTRUCTURAL SOUTH

TECHNO-ENVIRONMENTS OF THE
THIRD WAVE OF URBANIZATION

JONATHAN SILVER

THE MIT PRESS   CAMBRIDGE, MASSACHUSETTS   LONDON, ENGLAND

The MIT Press would like to thank the anonymous peer reviewers who provided
comments on drafts of this book. The generous work of academic experts is essential for establishing the authority and quality of our publications. We acknowledge
with gratitude the contributions of these otherwise uncredited readers.

This book was set in Stone Serif and Avenir by Westchester Publishing Services.
Printed and bound in the United States of America.

Library of Congress Cataloging-in-Publication Data

Names: Silver, Jonathan (Urban geographer), author.
Title: The infrastructural South : techno-environments of the third wave of
    urbanization / Jonathan Silver.
Other titles: Infrastructures series.
Description: Cambridge, Massachusetts : The MIT Press, 2023. |
    Series: Infrastructures series | Includes bibliographical references
    and index.
Identifiers: LCCN 2023000208 (print) | LCCN 2023000209 (ebook) |
    ISBN 9780262546874 (paperback) | ISBN 9780262376730 (epub) |
    ISBN 9780262376747 (pdf)
Subjects: LCSH: Urbanization—Africa—21st century. | Infrastructure
    (Economics)—Africa. | City planning—Africa. | Sociology,
    Urban—Africa.
Classification: LCC HT384.A35 S558 2023 (print) | LCC HT384.A35
    (ebook) | DDC 307.76096—dc23/eng/20230110
LC record available at https://lccn.loc.gov/2023000208
LC ebook record available at https://lccn.loc.gov/2023000209

10 9 8 7 6 5 4 3 2 1

For my dearest Helen and Uma

CONTENTS

ACKNOWLEDGMENTS

There are many people to thank for this book. The research and writing took place over a long period, and throughout these years, I've been lucky enough to spend time and learn from colleagues, friends, collaborators and research participants who have shaped my career and the opportunities that have opened up to me to write this. To have had the chance to spend my time walking and talking in so many different urban spaces and city places with so many wonderful people has genuinely been one of the best things in life. This book owes everything to those walks in the endless city—long may they continue.

I start with those at the Urban Institute at Sheffield where I am based and particularly Simon in shaping my career so much and who was joined by Aidan in pushing me to turn an idea of doing a book into a viable project and later Beth for helping me over the line and being so supportive when I really needed it. The debt of gratitude to Maliq, Michele, and Vanesa should be clear in the book. Your ideas and encouragements are always inspiring and it has been amazing to learn so much from you. And to Vicky S and Vicky H, Rachel, Andy, Rowland, Sam, Tom, Lindsay, Linda, Hita, Erika, Enora, Bert, Miguel, Alex, Tanzil, Lorraine, Matt, John, Ryan, Jenny, Paula, Victoria, and Glyn—all amazing colleagues at Sheffield who have pushed me along with the book. And recently to new friends in the corridor team—Fatima and Yannis—and the centripetal team—Adam and Rich—for our work together.

A meeting at the African Centre for Cities with Henrik and Mary during my PhD has been integral to this project and indeed so much of the underpinning of this book. Your kindness and support are immeasurable and your contributions to my thinking and academic (and non-academic) life vast. Later on, the Situated UPE group, a loose collective that formed and met at various times, has guided me greatly over the years and given me the opportunity to meet some awesome people with whom the walking and talking in various cities never stopped: Amiel, Wangui, Prince, Basil, Kathleen, Anesu, Cecilla, Alejandro and many others.

Alan/Renee, my North American friends and the times we've spent full of talks and walks, cycling, research and cats on both sides of the Atlantic—a constant in my life in so many ways. Omar who has been integral to how I come to think infrastructure in our thoughtful dialogue in which I always admire your original thinking as we walked and walked with Barbara from Detroit to Brussels and Ramallah to London.

At Durham University, I owe so much to Harriet and Cheryl for giving me an opportunity to become an academic through my PhD. Your support and guidance were (and remain) a model of generosity, care, and intellectual openness that made me feel so lucky and privileged. Colin for so much support, collaboration, generosity, encouragement (and banter) that it is hard to measure in any other way except the gold standard in being the best especially for our fieldwork together in Cape Town and Kampala. And to others in the geography department such as Gordon, Ben and particularly Marcus, who went from inspiring me as an undergraduate many year earlier in my thinking about Africa to amazingly becoming a colleague. Thanks also to Jayne, Ankit, Stella, Cat, JJ, Anne, Andres and others that I shared my PhD journey with.

In Cape Town at UCT and especially the ACC, Sue and Edgar, Maryam, Zarina, Suraya and Andrea, Nancy, Tau, Vanesa, Warren, Anton, Anna, Mercey, Liza, James, Katherine, and many others have been so kind and generous over the years. Thank you for being an open and inquisitive space. I learnt so much from the Sustainable Energy Africa team, including Melu, Mark, Megan, Yachika, and Simisha, all of whom have been great teachers. There are also people such as Sarah, Jacques and Cindy, Andrew, and others at the City of Cape Town to whom I owe much gratitude in our interactions. The folks such as Mark and Blake at the Sustainability Institute have always been kind and informative.

In Kampala, Joel has been a brilliant research collaborator and friend with whom to learn the city over the years and likely long into the future. To Sydney and our two-decade friendship that started in the village before moving to the town. I've always looked forward to hanging out and learning from you. At the Urban Action Lab Shuaib, Peter, Gloria, Hakimu, Disan, and Teddy as well as others remain such an inspiring group of scholars working in the city. My time speaking and working with architects such as Josephine, Alex, Mark, Doreen and their students opened up new perspectives. The residents of Namuwongo have always been welcoming and taught me more about the world than I could ever learn in a book.

In Accra, Christopher, whom I knew from Manchester, opened his door to me and the city. David and Kathleen gave me a roof over my head and lots of architectural education, as well as the chance to hang out on our street in Osu with people such as Sammy. Many thanks to the geography department for letting me visit during my PhD, including the guidance from Professor Jacob Songsore on Accra. And to Professor Brew Hammond for letting me spend time in Kumasi and giving me a crash course on Ghanaian energy issues. I owe a debt of gratitude to Simon and Innocent at ISSER for helping me think about infrastructure from a different part of the city. During my time in Ga Mashie, Solomon was an invaluable guide and walking friend, and like in Namuwongo members of the community were kind and generous in their time and insights.

My time at LSE Cities at the London School of Economics, working as part of Austin's project with Sobie, Kavita, and Astrid, meant so much to me after completing my PhD, giving me the confidence and intellectual excitement to take me into new research directions. The wider team at the centre meant always enjoying my trips down to London (and Delhi) and being part of the wider research culture, thanks to Adam, Suzy, Phillip, Catrina, Andrew, Ricky, Thomas, and Claire.

In Manchester, I've been lucky to hang out with a great set of academics and friends in my home city, where I undertook my MA and subequently with whom I've also often shared a cold picket line and participated in debates around African cities and UPE. Thanks to Seth and Simin, Nate,

Tom and Kate, Erik, Martin, Mark, Chris, Maria, Kevin, Christina, Diana, and so many others. Dan, my brother, and Ash, my best friend have also been ever present in my thinking and learning from Manchester, as well as always being there when I've needed help and support.

I have also been lucky enough to learn and discuss infrastructure with some great European colleagues over various inspiring events, especially at LATTS with Olivier, Syvly, Jonathan, Francesca, and others who welcomed me as a PhD student and who continue to support and encourage me. Also special thanks to Jochen and Sophie—I am so happy to have been able to spend time with you, learn from you, and have fun wherever we meet.

A shout out to the "Landscapes" group—Patrick, John, Anthony, Sarah, Sophie, Rich, Dillon, and Zac (as well as Alan and Nate)—my digital geography gang whom I always look forward to hanging out with in real life but also everyday online.

I would also like to acknowledge various funders that have allowed me to undertake this work: the ESRC for my PhD studentship and later for the SAMSET project, the Leverhulme Trust for my ECR fellowship that provided much of the space and time to research and write this book, and Formas for the HICCUP project in Uganda and also time for writing after my PhD. I have been fortunate enough to receive workshop awards from the Antipode, Urban Studies and Urban Geography journals that allowed me to develop my thinking further on infrastructure. The team at the MIT Press have been great, supportive and encouraging, especially Justin. Catherine's work on the index was exciting and helped me to think differently about the book.

Finally, I'd like to thank my family—mum, dad, Nickie, (Dan again), Rachel, Andy, James, Isaac, Jean, Robbie, as well as the London gang and those no longer with us, particulary my Grandad, Don who inspired so much intellectual curiosity. Thank you all for being so great. I surely would not have written this without your love and care over the years. This book owes most to Helen and Uma whose love and support made it all possible, and for the fun times we share that light up my life. Thank you.

1

INTRODUCTION: INTO THE INFRA-FUTURE

INFRASTRUCTURE ELSEWHERE

Pumzi, whose title means "breath" in Swahili, by Kenyan filmmaker Wanuri Kahiu, explores a future in which environmental collapse has precipitated a new world for the Maitu community in the East African Territory. A war over water has led to an irradiated landscape, and survivors live in extreme resource scarcity through an infrastructurally controlled, enclosed environment in which "the outside is dead." Everyday life in the community is powered through manually generated energy and produces zero pollution. Water is rationed, with sweat and urine recycled. Glimpses of a futuristic, high-tech city are visible through the windows. At one point in the film, the protagonist, Asha, escapes and watches the disposal of trash from the city into a landfill, exposing the deception of the "no-waste" system. *Pumzi* conveys both hope and dismay regarding the infrastructural futures of Africa. Through the psychological, social, and political costs to people of sustaining these networks, the film expresses the inherent limitations in the promise of infrastructure to maintain life in the city, the untamed natures of the "outside," and the myriad of technological and environmental futures that might open up.

Monsoons over the Moon, by another Kenyan, Dan Muchina, writing under the pseudonym Abstract Omega, also conveys a dystopian future in which authorities have shut the internet down. This, according to film critic Will Press (2015), clarifies that "Those who control the infrastructure for the storage and transmission of information eventually determine the content of the information itself." Mobilizing the brutal police violence experienced by Nairobi's youth in present-day neighborhoods such as Mathare (Kimari 2020), Muchina shows how the protagonist, Shiro, navigates the streets and technologies and the people and places of an alternative, futuristic Nairobi—a future that may yet come to pass. *Monsoons over the Moon* speculates on ways through which new technologies can create a repressive terrain and unequal experience of the city, while still leaving room for the capacity and agency of the human—in this case, Shiro—to make sense of, navigate, contest, and even transform these conditions.

These two African science-fiction films evoke infrastructure that exists both within and outside the contemporary moment. We can interpret *Pumzi* and *Monsoons over the Moon* as visions of futurity that rupture linear narratives of how these networked geographies have and will shape urban worlds. In doing so, they might prompt the viewer to reassess assumptions concerning the role of infrastructure across rapidly urbanizing Africa. These are presents and futures that remain simultaneously comparable to and highly differentiated from the experiences of what has come before, suggestive of the need for new ways of thinking about the fundamental importance of these systems in shaping urbanization.

This book sets out to address a critical question for our planetary urban futures: How do we account for the role of infrastructure in the urbanization of Africa? It is estimated that between 2010 and 2050, the share of global urban dwellers living in Africa will rise from 11 percent to more than 20 percent, as its annual urban population growth is sustained at nearly double the world average (CSIS 2018). With these urban transformations come massively expanding, intensifying, and contested infrastructural spaces, relations, and geographies, and a series of critical concerns. How will governing these urban systems proceed in towns and cities, doubling, trebling, or even quadrupling in size in a decade? What does it mean to center infrastructure in this massive change? How will this explosive urban growth intersect with the ways in which infrastructures are already entangled in historic and contemporary environmental, social, and economic crises? What does such rapid urbanization mean for how infrastructures shape everyday social reproduction and forms of inequality? And how different will this urban infrastructural experience be from what has come before? The central argument of *The Infrastructural South* is that we need new ways to explain the everyday functioning of basic services, unfolding hi-tech enclaves, new transnational trade corridors, and digital apps operating across urban space. This is because the uncertain futures of urban Africa are, as *Pumzi* and *Monsoons over the Moon* allude to, being written through ever-shifting techno-environments—that which surrounds and suffuses the making, operating, and experiences of infrastructure.

Take for instance Mbale, a town in eastern Uganda, which I have been visiting since 2003. For many years, there had been little maintenance and repair of its road network since it was built in 1954, and according to a local engineer, it has required comprehensive reconstruction since the 1970s. However, in 2013, a new program of investment beckoned. After receiving World Bank loan-based finance through the Uganda Support to Municipal Infrastructure Development fund, the main route through the center was upturned. Old trees were cut down; dust, sand, and concrete blew into people's faces along the now bright orange-colored road. There was much disruption, but it was a sign that something was happening, and better still, local people were being employed in the construction. In the meantime, there were traffic jams, with exhaust fumes suffocating everyone around, bored-looking faces peering out of windows of the backed-up matatus, endless noise from the heavy machinery, and piles of construction garbage piled up along the sidewalks, but it was understood that this was temporary and necessary. However, things did not go to plan. The road contractor, Zambia-based PTS, had built the road too high. When it rained, water spilled into the business premises along the side of the road. It was a disaster for local traders and needed entirely rebuilding. The road construction contract was transferred to a Chinese company, Zhougme Engineering Group. I couldn't believe, when on another trip to Mbale, that it still wasn't finished. It seemed like it might never happen. However, eventually, the road opened in 2019—five years late. A huge wait for many in the town. And yet by 2021, sections of the road, supposed to last fifty years, such as at Nabuyonga Rise were again covered in potholes. This story of investment in infrastructure, regimes of maintenance and repair, as well as the struggles to keep up with growing road traffic are familiar in many towns and cities. It is a tale of politics and materialities, local expectations and international connections, waiting around and navigating through, local businesses and construction contracts, finding employment and losing trade, improvements and backtracking.

To think about transport infrastructure in urban Uganda only through this lens of problem, failure, and disruption is to miss something, despite it being necessary. Across Kampala, a new set of digital apps and tools are being developed that open up possibilities to navigate the city better through various forms of transport such as the boda boda or matatu—and it is primarily being driven by young, tech-savvy people. In these quickly changing techno-environments, the youth are developing the capacity to navigate dense topographies that are simultaneously material and digital, and in doing so, they find ways both to inhabit the gridlocked, potholed city and to enhance livelihood opportunities by finding ways around, over, and under these blockages. Indeed, as Ronald, a tech entrepreneur whom we will visit again in chapter 8, recognized, young people are able to think about and act upon the rhythms and flows of everyday urban life through the in-the-making infrastructures they encounter. He said, "Young people are versatile, they like trying out new things . . . and they are fast." What begins as a problem of infrastructure under-capacity, breakdown, disruption, or lack of maintenance becomes a basis for a new kind of system being deployed across the city—ways to steer clear of dead ends and delays. Taking diversions sometimes opens up opportunities for digitally enabled hustles, side projects, and chances to make good. These opportunities for the tech literate emerge in ways that suggest what Tatiana Thieme (2013, 389) termed "alternative modes of social and economic organization amongst youth." Understandings of infrastructure in urban Africa, then, must find approaches that can hold together such tensions of promise and problem, future and past, big structural transformations and mundane, everyday inhabitation.

It is these ever-shifting networked geographies that do so much to shape urbanization and its experience by hundreds of millions of urban dwellers. To interrogate the infrastructure of cities such as Accra or Kampala requires thinking anew in ways that can capture a new set of coordinates. It means to move beyond the knowledge, practices, and assumptions tied to a Eurocentric idea of urban modernity and the city—an idea based predominantly on the technological experiences of the Western metropolis. In doing so, we encounter an urban history of the metropole that became simplified and then universalized and abstracted into a problematic way of understanding infrastructure elsewhere, the so-called periphery or South. This book proposes a different telling to move away from these traditions and into the future based on the Infrastructural South as both a condition and epistemology.

THE THIRD WAVE: AFRICA'S URBANIZATION

A SHIFT TO MAJORITY URBAN POPULATIONS

We are entering a new phase of planetary history: accelerating climate crisis and environmental collapse, rapid integration of advanced technologies into everyday life, the intensification of transnational migrations, 272 million people globally in 2020 (United Nations 2020), and intense social crisis, precarity, surging wealth, and glaring inequality nearly everywhere. The urbanization of Africa arguably forms a critical locus of this era. In effect, this third wave of urbanization exceeds in speed and intensity the growth of Western cities, or the modernization period in newly independent nation states and emerging economic powers in the South from the 1960s. Population data from UN-DESA (2018) helps to convey sub-Saharan Africa and Southern Asia as the key crucibles of this transformation. By 1950, the so-called Global North (comprising Europe with 51.7 percent, North America with 63.9 percent, and Oceania with 62 percent) was majority urban, reflecting the growth of cities intimately tied to industrial capitalism that expanded across the nineteenth and twentieth centuries. Between 1950 and 2020, the urban and population growth of what was termed the Third World saw other regions become majority urban.[1] This too was accompanied by industrial growth aided by modernization programs. Today, the only continent that remains with a majority rural population is Africa, not estimated to become majority urban until 2034.

These regional figures belie subregional variations, with Southern Asia not projected to become majority urban until 2044, whereas Southern Africa became a majority urban subregion in 1993, Northern Africa in 2008, and Central Africa in 2019, with West Africa not estimated to attain this status until 2024, and East Africa still projected to be majority rural in

2050, even with rapid urban population growth taking the proportion to 47 percent (UN-DESA 2018). However, overall projections give credence to the claim that the future centers of urbanization are geographically concentrated on sub-Saharan Africa and Southern Asia.

SPEED AND INTENSITY

We can understand Africa's place within this third wave of urbanization not simply as a shift to majority urban populations but by the sheer numbers of people involved in this transformation and at such dizzying speed. In 1950, Africa had one of the lowest urban populations, with only 32.5 million people living in towns or cities.[2] We can compare this to 284 million in Europe, 110 million in Northern America, and 246 million in Asia. By 2050, Africa will make up the second largest total urban population with nearly 1.5 billion urban dwellers, dwarfing Europe (600 million) and Northern America (386 million) combined by hundreds of millions of people. The scale and pace of this transformation, only matched by Southern Asia in the coming decades, mean that despite the huge numbers of people living through this experience, much of the urban space and accompanying infrastructure required for these surging populations have yet to be fully established, planned, or built.

It is worth inspecting these population dynamics in sub-Saharan Africa in more detail. In the last twenty years alone, more than 200 million more people in the region call the city (or town) home. With about 1.5 billion urban dwellers by 2050, the current topography of existing cities is likely to be significantly altered to accommodate this population growth. And yet, urban Africa will, according to many indicators, remain "undeveloped" and "in crisis" compared to the cities of the North and, increasingly, the East. And this pace of change is set to continue well into the twenty-first century, with UN-Habitat (2014, 16) estimating, "The total population of the continent is projected to nearly double from around one billion in 2010 to almost to two billion by 2040 and may well surpass three billion by 2070" when the "continent is predicted to be more than 50 per cent urbanized."

The average urban growth rate in the region is 3.5 percent per year. This compares with a global average of 1.63 percent per year between 2020 and 2025. The population projections of large metropolitan areas that dominate representations of urbanization in the South help to convey the magnitude of this transformation. Forecasts project Lagos to grow from 7 million in 2000 to 18 million by 2025, Kinshasa from 5 million to 14.5 million, and Nairobi from 2.6 million to 6.1 million (UN-Habitat 2014)—a quarter of a century for urban populations to triple in size. Projections to 2100 suggest that sub-Saharan Africa may contain the three largest global cities—Lagos (88.3 million), Kinshasa (83.5 million), and Dar es Salaam (73.7 million)—and thirteen of the top twenty most populated cities, all more than 35 million (Hoornweg and Pope 2017; see also Satterthwaite 2017). If there remains uncertainty about these statistical estimates, they offer a snapshot of Africa's rapid urbanization and its likely continuation over coming decades. As I set out above, the 1950 total urban population in Africa was a mere 32 million—a figure that, extraordinarily by 2100, could correlate to just one metropolitan region.

However, it is not just the spatial form of the megacity or large metropolitan region through which this urbanization is proceeding apiece. Urban Africa is also being assembled across many small towns and cities that are "off the map" (J. Robinson 2002), rarely receiving much scholarly, policy, or public attention. In these urban spaces of fewer than 500,000 people and accounting for up to 50 percent of the urbanization in the region (Pieterse and Parnell 2014; see also Hardoy and Satterthwaite 2019), populations can double in the space of a decade. These urban revolutions are leaving under-resourced municipalities with little scope to plan, deliver

1.1   The eastern Uganda town of Mbale.

infrastructure, or support new and growing communities. For instance, the population of Mbale has been predicted to grow from 50,000 to 150,000 between 2002 and 2022 (Mbale Municipality 2010)—a tripling of the population in less than two decades.

URBANIZATION WITHOUT INDUSTRIALIZATION

Another factor for arguing for the need for infrastructural perspectives in this third wave is the disconnection between urban growth and industrialization in which "many of today's developing countries have high rates of urbanization with little significant industry" (Gollin 2018, 36). If we consider manufacturing as the central activity of industry, then much of the urbanization beyond the West from 1950 onward was connected to modernization programs and the emergence of the factory and planned industrial zones in the city. However, in recent years, this has not been the case: the industrialization process has either stalled or never got properly started. Rather, the economies in these countries are often dominated by two sectors.

First, extractive enterprises incorporating various resources from mining to timber through to cobalt are located beyond urban boundaries, even as they are integrated into a system of planetary urbanization (Arboleda 2020). This has meant that "resource-exporting countries may urbanize without acquiring the industrial sectors that we typically associate with development" (Gollin 2018, 38), leaving an ever-greater number of people locked out of formal employment. Such dynamics chime with my experience in Uganda. For nearly twenty years, I have listened (and indeed aged considerably) to President Museveni talk on the radio about the need to develop processing and manufacturing capacity—still without success and with ever-growing urban populations and youth unemployment running high.

A second economic sector has also emerged as a relatively recent and intensifying dynamic: the transformation of urban land into development projects on the "real estate frontier" (Gillespie 2020; Obeng-Odoom 2015a These circulations of regional and global finance into the built environment proceed with little correlation to industrialization or industrial sector employment. Real-estate developments and "new" cities (Ehwi and

Morrison 2022; Olajide and Lawanson 2022) emerge at a rapid and not so rapid pace. Such speculations take place amid the ever-present incremental, self-built housing projects across many urban spaces. These are especially prevalent on the periphery (Izar 2022; Meth et al. 2021) of large metropolitan regions, such as Kasoa, which sits just outside the Accra region—a landscape of half-built homes, concrete foundations, security guards, fragile ecologies, big gates, and newly established congregations meeting in vast places of worship anticipating future growth. And throughout these spaces, promised and failed infrastructure projects dot the landscapes: half-built roads replete with slow moving machinery, concrete flyovers soaring into the air, new but not entirely useful solar-powered streetlights, signs, and boards, some more faded than others proclaiming future opportunities and investment. Amid this urbanization, there appears little correlation between any kind of industrial activity and the ever-incessant urban growth, despite a scattering of new logistical warehouses or perhaps a salesroom spilling into the road with rows of new construction equipment.

Intimately tied to this urbanization without industrialization and especially the incremental (re)making of the city are the resource-poor neighborhoods in many diverse spatial forms that span so-called formal and informal conditions and often located both physically and in other ways in highly marginal and contested spaces (Brown-Luthango 2021; Ngwenya and Cirolia 2021). The slum, perhaps better understood as the popular neighborhood, has always been a key feature of urban modernity, all the way back to the industrial cities of northern England through to the squatter settlements of the colonial era, the rapid growth of post-independent urban areas and the struggles for survival, and neoliberal assault during the years of structural adjustment programs. With more than 50 percent of all urban residents in sub-Saharan Africa living in "slums," estimated at 238 million people (UN-Habitat 2020), this might rise to between 800 million and 1 billion people by 2050. If we characterize these spaces through the need for urban dwellers to navigate the lack of access to basic services (Dakyaga, Schramm, and Kyessi 2022), demands of repair and maintenance (Wahby 2021; Jambadu, Monstadt, and Schramm 2022), security of tenure (Uwayezu and De Vries 2018), and exposure to risk and uncertainty (Adelekan 2020), we should never reduce or flatten them simply to such deficits or gaps. People find ways to inhabit the infrastructures of these popular neighborhoods in ways that open up, extend, connect, and contribute to the everyday lives of these communities. In these spaces, infrastructure is both a site of struggle and survival, and a basis for envisaging and making real speculations for learning, improvement, celebration, and collaboration.

It might not be possible to comprehend, think through, and decipher this urbanization process in any kind of systemic way, given its size, scope, and diversity. There has, however, been inspiring work articulating the various dimensions of urbanization across Africa that open up important perspectives and entry points (e.g., F. Demissie 2012; Guma 2020; Hakam 2019; Lawanson, Salau, and Yadua 2013; Lwasa 2014; Myers 2011; Pieterse 2013; Scheba 2021; Simone and Pieterse 2018). These interventions have attempted to think through urban Africa via multiple different threads, narratives, standpoints, and propositions from associational life to governance, city life to violence, livability to religion, mobile technologies to environmental change. However, I would contend there remains the need for an integrated rethinking of infrastructure within and across this process, another line of thinking into the massive, world-changing shift to an urbanized society that revolve around and spin out of Africa's urbanization and the uncertain futures it generates (Karuri-Sebina, Haegeman, and Ratanawaraha 2016).

THE INFRASTRUCTURAL SOUTH

THINKING URBANIZATION WITH AND THROUGH INFRASTRUCTURE

The infrastructural turn is now well established in the social sciences (Graham and Marvin 2001; Star 1999). However, it is briefly worth setting out why these urban networks matter. Infrastructure is a complex and interconnected entanglement, facilitating the flows and circulations required for everyday social reproduction for households such as water and light. These systems also enable economic exchange such as the fiber-optic cables of 4G and 5G networks needed for instantaneous global trade. Various actors deploy these systems across a variety of geographical scales operating and connecting the home, neighborhood, city, region, and beyond, yet often "giving no hint to the average user of the huge and geographically stretched infrastructural complexes that invisibly sustains them" (Graham 2009, 6). Infrastructure is important not just as the "fundamental background to modern everyday life" (Graham 2009, 1) but "can be conceptualised as a series of interconnecting life-support systems" (Gandy 2005, 28) for cities through the urban service provisions that sustain the town or city.

Infrastructure offers a powerful device in which to understand the urbanization process as characterized by constant flux and movement that enable a series of actors to navigate, shape, and live through ever-shifting geographies. The book is an attempt at a new opening in thinking about the dynamic role of infrastructure in urbanization—one based on the organizing concept of the Infrastructural South as both condition and epistemology.

INFRASTRUCTURAL SOUTH AS CONDITION

The Infrastructural South as condition demonstrates the need to rethink how techno-environments have shaped historical and contemporary urbanization across Africa. It denotes the material geographies that we encounter across towns and cities when we think about infrastructure. Current assessments suggest a significant infrastructure "gap" for the region compared to what exists elsewhere. As UN-Habitat (2014, 20) argued, "African countries and cities are burdened by high infrastructure deficits and shortages in access to technologies and services." Estimates by the African Development Bank (2018) propose that the region's infrastructure needs amount to USD 130–170 billion a year, with a financing gap in the range of USD 68–USD 108 billion. This deficit exists even as surging flows of investment into new transportation and development finance profoundly reshape urban spaces from Accra to Zanzibar City. Planners justify such infrastructure projects as aspirations to close the productivity gap with other world regions, to improve economic performance, and to stimulate new enterprises and employment opportunities as national economies and growing urban populations demand. National and regional-led infrastructure development plans closely entwine them with megaprojects of dams and railways, ports and export zones, most notably these days through new circuits of Chinese loan finance (Asante and Helbrecht 2020; Goodfellow 2020). In doing so, these infrastructures reconfigure extractive hinterlands, global connections, and regional aspirations as finance shapes elite enclaves and technologies of accumulation for the few.

This infrastructural gap also pertains to essential urban services such as water, sanitation, and electricity that sustain everyday social reproduction, with varying conditions across and within different cities, regions, and the continent as a whole. For instance, only a third of urban populations have piped water access into the home—a reduction since 1990 when 43 percent of households had access (Satterthwaite 2017). And only 50 percent are using safely managed drinking water services (including piped but also other sources), with figures as low as 16 percent in Uganda (World Bank, n.d.). Access to electricity by urban populations was at 78 percent in 2020 (World Bank n.d.), with some significant variations (e.g., Uganda 69 percent of urban population, Ghana 94 percent, and South Africa 88 percent). The data for access to clean fuels and technologies for cooking show the challenges of providing energy that doesn't damage household health or surrounding environments—currently only 14 percent of households in the region. Waste as a proportion of municipal solid waste collected increased in sub-Saharan Africa from 32 percent to 52 percent between 2001–2010 and 2010–2018 (World Bank n.d.), but these statistics highlight the significant task facing people without formal waste-collection services, especially in relation to the environment and pollution, alongside a cascading health crisis.

The need to deploy and operate various infrastructures, both economic and social, has always been a critical aim of what we know as "development." In recent years, development officials and organizations have mobilized these imperatives and subsequently operationalized and measured them through the Millennium Development Goals (MDGs) and the more recent Sustainable Development Goals (SDGs; Sachs 2012), which have 17 goals, 169 targets, and more than 300 indicators, many of which pertain to infrastructure. Up to 50 percent of required investment to achieve the SDGs will need to be focused on the deployment of various infrastructure services (Adshead et al. 2019). This both reflects the primacy of infrastructure to these broader developmental programs and is suggestive of a way of thinking in which these infrastructural challenges faced by governments and populations can be "solved" based on the experience of "developed" Western cities. However, a growing number of scholars and activists have

1.2 Accra, a city long the target of development.

criticised the technocratic, top-down nature of the MDGs and SDGs and how the powerful from countries in the North to corporations to unaccountable nongovernmental organizations (NGOs) have shaped the underlying ideas, logic, and outcomes (Esquive 2016). Samir Amin (2006, 1) questioned "whether they are mainly ideological cover (or worse) for neoliberal initiatives." And Jane Briant Carant (2017, 39) argued, "As long as the goals remain steeped in power-laden hegemonic frameworks, serving only as an opportunistic medium through which power interests can assert, maintain and defend their position and preferred economic modalities, poverty eradication will remain relegated to the imagination."

Clearly then, infrastructure is a prerequisite of achieving the SDGs and yet is something that cannot be organized through these types of technocratic fixes predicated on ever more loan-based finance and forms of often imported technical expertise. Instead, such aspirations may need a revised understanding of the geographies of infrastructure in the urbanization process—one that is able to shift beyond the problematic logics of development and the SDGs. The need to center the political dimensions in the making and operation of infrastructure, then, is urgent, particularly as up to two-thirds of the investment in these systems up to 2050 remains to be built. The infrastructural stakes are high in urban Africa. From the basic services required for everyday life (Amankwaa and Gough 2022; Narsiah 2011) to the infrastructures of mobility (Nyamai and Schramm 2022; Ibrahim and Bize 2018) to the global technologies of exchange and accumulation (Kilaka and Bachmann 2021; Githaiga and Bing 2019), unwritten futures are coming into view. How it is shaped will be determined by the power geometries and urban politics that influence the planning, deployment, and operations of these networks, as well as the multitude of tactics, strategies, and calculations in which people find new ways to inhabit the infrastructural city.

POLITICIZING THE TECHNOLOGIES AND ENVIRONMENTS OF THE INFRASTRUCTURAL SOUTH

Underpinning the writing of this book is the acknowledgment of the need to politicize infrastructure to demonstrate that the Infrastructural South as condition—that is, its material geography—is always shaped by the politics and power relations surrounding these systems (Graham and Marvin 2001; Lemanski 2020). Infrastructure is not a neutral set of technologies. Rather, actors, ideologies, and social relations imbue it with the power and politics of the world around (Desai, McFarlane, and Graham 2015; Mkutu, Müller-Koné, and Owino 2021; von Schnitzler 2013). Scholars have explored the deeply political nature of infrastructure across multiple different sectors, including water (Truelove 2011; Millington and Scheba 2021; Musemwa 2010; Tchuwa et al. 2018), sanitation (Arefin 2019; Chaplin 1999; Nakyagaba et al. 2021; Swanson 1977), and electricity (Baptista 2015; Guma, Monstadt, and Schramm 2022; Jaglin 2008). These studies have rightly argued that infrastructure reflects and reinforces wider social and political relations across urban worlds.

Let us turn to the writing of South African novelist Niq Mhlongo in *After Tears* to consider this infrastructure politics in more textual detail. Mhlongo (2011, 156) writes:

I recalled that the switches were dead when I had come back from Naturena that morning. At the time I had thought it was about the usual problem with the old power station in Orlando East.

"Siriyasi, Advo, these capitalists have removed the cables. Ever since we voted for them they don't give a fuck about us any more," said Zero, anger registering in his face.

"They claim that we are stealing electricity. To get reconnected we need to pay one thousand five hundred bucks. That's why there's an urgent meeting today. The residents are angry, Advo. I've never seen people as angry with the government."

The passage offers a useful snapshot of the politicized, unequal geographies in the delivery of services in South Africa. It focuses on the anger of two of the main protagonists concerning the disconnection of electricity in their neighborhood and the mobilization of the community to challenge the utility company. The disconnection provides an example of infrastructural inequality and shows how access or not to these networks can negatively impact on urban dwellers (Jambadu, Dongzagla, and Kabange 2022). It is suggestive of decisions about the planning, distribution, and operation of infrastructure as always political, sometimes contested and often controversial. Infrastructure becomes the material site of both the expression and contestation of power over the urban and its future. This framing of infrastructure as political is in contrast to much of the developmental, planning, and policy literature of African cities, which instead focuses on technical and managerial responses.

The unequal geographies of infrastructure and the cumulative consequences of failures to provide essential service provision were made clear to me during fieldwork for my doctorate in Cape Town. On a bitterly cold day in the Western Cape, the rain was hammering down on the corrugated iron roof of a "low-income" Reconstruction and Development Programme (RDP) house.[3] I was undertaking household surveys and interviews, and "Alice" kindly me invited inside. I felt the damp immediately. The condensation dripped down the walls and the moisture, combined with the cold, made a rather unpleasant microclimate as I stood there wrapped up in my coat. In what must have been the bedroom, I heard a child coughing and crying. One of the family told me about the illnesses and sickness they had suffered that winter. Colds, pneumonia, and flu had made it a difficult time of year, but they were thankful that tuberculosis hadn't visited that winter season as it had in some other households in the area. Before I left, one member of the household caught me staring at a heater that looked like it could make some inroads into the icy chill that pervaded the home

1.3   RDP houses in Mamre, Cape Town.

and said, "We cannot afford such luxuries as electricity at this time of the month."

This experience brought home the everyday injustice and struggles that are wrapped up with infrastructure; even if the technological arti-facts existed for access (i.e., the electricity connection, the heater), it did not mean that the household were accruing any benefit from it if they could not afford to operate it. I left feeling not only upset but also angry and determined to develop my skills and knowledge to understand the centrality of these systems better in the making of urban inequality and the politics that produced such conditions or indeed might transform these situations. To do so, I realized, meant a revised way of understand-ing infrastructure, particularly as someone who had taken services such as electricity and water for granted for much of my life in the UK. The experi-ence opened up an ongoing process of unlearning many of my (Northern) assumptions about the urban worlds constituted through these networks.

INFRASTRUCTURAL SOUTH AS EPISTEMOLOGY

As I've set out above, the term "Infrastructural South" denotes a geography or condition beyond the Western city. However, I also want to emphasize that I use the term as an epistemological position to decenter Western narratives of urban modernity. This is a key aim of the book: to develop a postcolonial urban theory of infrastructure that might better comprehend and therefore politicize Africa's urbanization.

My position has developed out of a decade-long series of collaborative conversations and debates, investigations, and experiences examining the urban question. At the center of this shared journey has been the intention to think through Gayatri Spivak's (1993) provocation to unlearn our assumptions about knowledge and our own positionalities as scholars. This meant, in the words of Dipesh Chakrabarty (2008), to "provincialize" infrastructure, to open it up further to postcolonial debates and situated imperatives. Thinkers such as Ananya Roy (2009) have inspired me and colleagues to think about "the 'Global South' as an epistemological location—rather than a geographical container—through which a provincialization of dominating theory can be crafted" (Lawhon, Ernstson, and Silver 2014, 505). I borrow from and hopefully advance this collective work in proposing a related mode of theorization in thinking through the Infrastructural South—one that denotes an epistemological space to think anew about how we understand the third wave of urbanization and specifically the critical role of infrastructure.

The departure point for this book is in recognizing that we now live in an era that exceeds the already-unsatisfactory modes of thinking concerned with urbanization in the North, or indeed the post-independent modernizations of the South later in the twentieth century. I have suggested that the third wave of urbanization is not following the trajectories of what has gone before as a singular, uniform process, but rather is configured as a distinct, dynamic, and hybrid geography of infrastructural modernity. If these conditions cannot be interpreted from these terrains, then it is clear we need new vocabularies, explanations, and theory.

It is worth reminding ourselves of the argument Jennifer Robinson (2002, 583) made two decades ago and how this challenge to urban studies requires ongoing work to address because "urban theory is based primarily on the experiences and histories of western cities—much as Chakrabarty suggests that the theories and categories of historical scholarship have been rooted in western experiences and their intellectual traditions." In order to think anew about the Infrastructural South as condition also means to think about the Infrastructural South as epistemology, to open up lines of enquiry focused on the techno-environments that shape patterns of urbanization. To undertake such a task means to begin again with the narratives of urban modernity that we have been told and in which infrastructure features prominently.

THINKING ANEW

The Infrastructural South, as an organizing anchor for this book, is put together through a series of propositional techno-environments that differentiate urban Africa from the geographies of the abstracted Western city beyond a series of negative reflections. To do so requires a renewed reading of urban modernity as a basis for the seven techno-environments I propose that makes clear the importance of thinking about Africa's urbanization in relation to but also exceeding the experience of modernity in the metropole. This involves acknowledging that urban modernity always mutates beyond supposed Eurocentric dimensions, teleological models of progress, and the accompanying indicators of control over nature through technological innovation. As African science-fiction films such as *Pumzi* and *Monsoons over the Moon* are so adept at communicating, a different set of coordinates encompassing the Infrastructural South offers the capacity to think anew about how knowledge is currently produced and how we can shift beyond inadequate scholarly traditions.

In undertaking this task, I bring into view assumptions about the spaces that have been prioritized and ignored in the narrative of modernity and therefore urbanization as derived from a Western vantage point (Ferguson 1999; Mitchell 2000; J. Robinson 2002). The telling of the histories and presents of urban modernity as open, unfinished, and plural become a necessary means to begin to piece together the idea of the Infrastructural South. The poetry of Ian Kiyingi Muddu (2015, 68) on Kampala is suggestive of this hybrid experience of urban modernity. He writes:

When, as Muteesa's capital, it dispensed the right to life and death.
Its soils Muganzirwaza washed with the
blood of her sons, dissidents, a city of two tales:
Of the castrated old,
And the still born of a modernity betrayed by a Makerere baked brown
countenance on a black torso mumbo jumbo.

Kampala is both and neither modern and unmodern, built upon extended histories and traditions, negotiations, and experiences while also facing the future, the new and yet to be made. Modernity finds expression within and beyond these apparent binary categories of modern/traditional in this hybrid city. To be clear, Kampala is as modern as any other city, despite its historical and contemporary representation as a not-yet-modern city. Here, I wish to foreground how understandings of urban modernity have long been derived from, and interpreted through, the type of infrastructure in operation and resulting techno-environments that surround and suffuse these networks. Urban modernity, I contend, has been associated

1.4 Kampala, the hybrid city.

with the supposed completed networked infrastructures of the Western city (Graham and Marvin 2001). This paradigm is universalized in how cities are understood. When seen in the mirror of modernity, the African city appears as the *un*modern city.

This making of urban Africa as unmodern through a technological reading does not simply negate existing presents and emerging futures but also erases urban infrastructural pasts. For instance, Benin City or Edo was the capital of the Kingdom of Benin with a history that spanned to AD 900, reaching the height of power by the twelfth century. The ruins of the city looted and burnt by the British in 1897 were still indicative of the wealth and civilization of Edo. Evidence provides examples of the precolonial use of engineering to order and control nature through town planning and urban infrastructural networks. The Kingdom of Benin could instigate vast engineering projects to tame nature required for the growth of the city. Ekhaguosa Aisien (in Acey 2012) drew on oral traditions to tell of a significant engineering feat in the late sixteenth century by master civil engineer, Osiezeka, who oversaw the draining of a swamp and channelling of water into the Ogba River for Oba Ehengbuda, with intention being to enable expansion of the city and address the health problems caused by adjacent swamps. Later, Benin City operated a network of streetlights fixed in place with metal lamps and using palm oil fuel to provide light for nighttime travels. Colonizers looted some of these infrastructural artifacts, which are now held by the British Museum. The supposed hallmarks of Western infrastructural modernity, the technologically enabled taming of nature and an operational infrastructural network (Kaika 2005) are evident in Benin City long before the British imposed colonial rule.

The way this book picks apart the Eurocentric, linear conception of modernity is through consideration of the paradoxes encased in the multiple infrastructures I have encountered across urban Africa and the ways these exceed modes of description from the experience of modernity elsewhere. Infrastructure is a critical site of what Marshall Berman (1983) termed "creative destruction" into the varied forms of urban life and the technologies deployed to control untamed natures across urban space. This comprehension of urban modernity shaped through technology is intimately connected to the production of knowledge of African cities as unmodern, primarily through what we know as the networked city model (Monstadt and Schramm 2017). The model has shaped the understanding of the infrastructure in operation (or not) across urban Africa as unfinished and incomplete, even as problematic ways of conceptualizing urban technological change suffuse this perspective. Vocabularies for analysis of urban Africa need to broaden both to incorporate and to step beyond this incomplete perspective to consider how this supposedly universal, teleological experience of infrastructure is actually in a state of constant flux and mutation. To do so echoes a broadly postcolonial orientation that challenges the linear assumptions of Western modernity and into the diverging time/space of urban Africa.

The networked city model was derived from a limited reading of the Western experience of infrastructural modernity that became abstracted and then universalized. Many of those living or researching these networks know that the urban geographies of the Infrastructural South are not following in a linear fashion the histories of these systems in the North. We know that the supposed uniformity of technical networks throughout Western cities is not substantiated amid the diverse, highly uneven infrastructural landscapes of urban Africa, both historically and in multiple presents and futures (Bakker 2003; Lawhon et al. 2018). We know that this heterogeneous, hybrid, and partially networked landscape exists, despite the power of the networked city model as the primary "circulating ideal of modernity" (Monstadt and Schramm 2017, 104). We know that the networked city model has long shaped the underlying logics of urban planning and service delivery for highly uneven colonial governance and subsequent postindependence rule by politicians, planners, and engineers (Nilsson 2016)—that it has driven the developmental, neoliberal management of cities in the contemporary era. And we know that infrastructure has long been a lens through which the development sector has measured the progress of African cities against a universal model of progress and subsequently understood as in a technological state of "incomplete modernity." In this book, I address what remains to be fully articulated: the multiple, overlapping, and ever-shifting techno-environments of the Infrastructural South both as geography/condition and as the basis for a theoretical orientation beyond existing traditions mired in problematic notions of the urban networks of the modern world.

APPROACHING THE INFRASTRUCTURAL SOUTH

In proposing the idea of the Infrastructural South, I have focused beyond one city to generate data and develop theorization. I have undertaken a comparative, relational mode of thinking about the techno-environments that surround and suffuse infrastructure, extending Jennifer Robinson's (2016) call to think cities from elsewhere. Through extensive fieldwork across a number of urban spaces, and through particular case studies oriented toward various systems, this approach has allowed for different types of thinking than a book built up through engagement with a singular "city." Here, I highlight a focus less on the cityness of my research, rather engaging with the role of infrastructure comparatively and as a broader mode of thinking that exceeds the container of a particular place. As Garth Myers (2018, 15) has argued about such an approach, "This is a comparativism that makes use of the thick descriptions of particularities in the global South by testing their relation to one another." Of course, this has limits, despite working for more than a decade in various towns and cities, with large periods of time spent undertaking fieldwork, often in teams of researchers that allow for wider understandings. In conducting the research, much of what was learned needed to be unlearned, through reading, listening, walking, and talking on the streets of various neighborhoods, across diverse infrastructural terrains, amid times of anger and tension, celebration, and most often the mundane rhythms and flows of everyday life.

The methodological approach also builds into a longer personal history of learning and unlearning the urban—one that started in a Global North city familiar to the story of urban theory: Manchester. I have sat and written part of this book in my home district of Ancoats, a place Engels (1892) explored with incredibility during his time in what some term the original modern city, notorious for its slum conditions, pandemics, violence, and William Blake's (2004) 1808 description of the "dark satanic mills" of "cotton capitalism" (Beckert 2015). I have followed how the making of modernity in my home city enveloped not just the working class of northern England but the precolonial kingdoms of Africa for labor, the indigenous people of the New World as their land turned to plantation, and the workers in a decimated Indian textile trade. It is an important anchor to acknowledge, and it is critical to how I have composed this book as a relational, ongoing conversation across different but connected urban spaces.

The research that underpins this book spans various individual and collective projects, often working with research assistants in particular places, such as Joel in Kampala or Sydney in Mbale. It has involved working as part of academic collaborations with Colin, Simon, Harriet, Henrik, Mary, Alan, Omar, and others. I have been involved in partnerships with NGOs such as ICLEI and Sustainable Energy Africa, large organizations such as UN-Habitat, various municipalities, as well as time with African students teaching and training, alongside organizing field trips for UK students to places such as the Sustainability Institute in Stellenbosch. It has involved spending much appreciated time at various university institutions such as the African Centre for Cities at UCT, Geography at Legon, and the Urban Action Lab at Makerere and seeing the dynamic learning that takes place in these spaces, as well as connecting up to and being inspired by young scholars in different cities that always push my thinking. I've been able to conduct dozens of interviews with community leaders, policy makers, municipal staff, social movements, NGOs, political parties, technicians, and utility workers among others. I organized and participated in workshops with resource-poor communities through to national planners. I conducted household surveys in Cape Town, Kampala, and Accra that provided another layer of understanding to the more qualitative research. These methods proceeded alongside ongoing ethnographic work such as participant observation—sometimes with community members in various contexts and other times accompanying maintenance workers across the infrastructural city or attending a policy conference. I've also spent time developing more creative ways of conducting research such as using participatory photography methods alongside young people or engaging in speculative design sessions with architecture students. I've always added to my primary research methods through detailed policy analysis, putting together databases of key trends and drawing on various secondary data sources.

Underpinning this myriad of assembled methods has been a commitment to walking the infrastructure that makes up urban space as a way of pulling together all of this work. This method has led me along railway tracks and electricity lines, through the social infrastructures of particular neighborhoods, out of the city to large-scale systems such as hydroelectric power plants and ports, back to the center, and out on the margins. It did not focus exclusively on the urban spaces that this book engages with but when the opportunity has presented itself, sometimes as part of associated research or workshops and at other times meeting with and learning from friends in places such as Doula, Yaoundé, Maroua, Timbuktu, Bamako, Saint-Louis, Bobo-Dioulasso, Cape Coast, Dakar, Dar es Salaam, Zanzibar City, Nairobi, Jinja, Kisumu, Johannesburg, Durban, Marrakech, Kumasi, and Mbabane among others. As Andy Merrifield (2002, 14) argued, "truth claims about cities must be conceived from the bottom upward, must be located and grounded in the street, in urban public space, in everyday life."

This unlearning/learning anew has also proceeded outside of towns and cities, including months of village life in my early twenties in eastern Uganda and extended trips into hinterlands and the small towns and villages that lay upon these routes. I began to understand the village as a foundation to everyday life and the city a place to take risks, follow opportunities, or simply sell one's agricultural produce at the end of growing season and return home. Some would stay a night, a week, or perhaps indefinitely in the city, but the village was the anchor of life. Times do change. Increasingly, young urban Africans have fewer links to the village of their families, and yet this experience of a mobile life or seeing urban Africa from the village stayed with me as I began to undertake my research. It provides another part of the toolkit that has helped to produce this book.

It is from this patchwork of experiences as well as the more focused research and time in Ghana, South Africa, and Uganda that I feel it important to make claims about an urban Africa, despite the risks of flattening its intense heterogeneity. First, I think it is important to recognize some commonalities as set out above in its (infrastructural) experience of urbanization. As Simone (2004a, 18) argued, "African cities have also historically found themselves in the same boat when it comes to piecing together a functional sense of coherence and visibility from a most haphazard collection of aspirations and livelihoods" and therefore become "the objects of specific policy and program initiatives and administrative functions that are organised along regional lines." Despite the material, political, and social differences, and the divergent ways of understanding urbanization processes, particularly across Anglo-Franco divides (Hakam 2019), there is evidently some utility in also drawing these together analytically. Second, as James Ferguson (2006, 5) makes clear, this analytical position is also politically important when thinking about Africa, given it is a "category that (like all categories) is historically and social constructed . . . but also a category that is real, that is imposed and according to which, people must live." As I will explore in the next chapter, this idea or category of Africa has had profound consequences in how its urban spaces have been understood, planned, and experienced. Being able to push back against these traditions is vital and politically necessary if Africa is, in Ferguson's words, "real" for global finance or for Western-dominated institutions such as the World Bank or IMF. We cannot leave it to them to define its meaning but rather we must develop counter interpretations, meaning, possibilities, and ultimately a politics that is both critical and propositional against these forces.

Ultimately, my positionality as an outsider and as a European has been a necessarily complicated and demanding one, always negotiating what the purpose of my research and proposed analysis have been, what interests they serve, and how I can move beyond a Eurocentric understanding that doesn't replicate problematic traditions of engagement. I've been helped in thinking these challenges through by conversations throughout the research and writing of this book. This is perhaps best conveyed though what Renee C. Neblett told me during a visit to the Kokrobitey Institute outside Accra. She explained that "the bottom line is Africa has been conquered, it's a colony, ideas, development, imagination are products of a big lie. [We] need an opportunity to re-imagine, dream, reflect and that's not a period the continent has enjoyed for a long, long time and it's still not there." In my work and in the writing of this book, these words resonated, giving me a motivation to unlearn and learn anew, to challenge the representations and knowledge of Africa and the African city as category in the world, and to help open up pathways, opportunities, and support for a new generation of young African researchers to shape thinking on urban infrastructure.

STRUCTURE

In chapter 2, I set out the argument for the Infrastructural South. I critique how we have come to understand modernity. I show how this translates into a Western-dominated narrative of urban modernity and its close association with the infrastructure that made the modern city. In the subsequent chapters, I explore how the Infrastructural South is produced through a series of techno-environments drawing on and advancing the ideas that have emerged out of the field of urban political ecology (UPE). I define a techno-environment as that which surrounds and suffuses infrastructure. In my work, I have found the usefulness of defining infrastructure more narrowly than others as the material systems themselves, whether transport or waste. Using the idea of the techno-environment therefore points to that which exceeds the material configurations of these networks, especially the politicized relations and processes that incorporate the social, technical, ecological, or economic dimensions that shape the planning, deployment, operation, and experience of, say, a sanitation system or energy network. For each of the techno-environments that I propose—enclave, incremental, imposition, corridor, digital, secondary, and predicament—I draw on case studies from across urban Africa and wider trends, alongside ongoing debates in urban (infrastructural) studies scholarship. This work builds toward an analysis focused on four analytical foci: *operations* that compose the techno-environment, incorporating the key features, logics, and how they operate across urban space; *geographies* that are being generated by the techno-environment, alongside how and why they diverge from the universalized, abstracted model of urban modernity; *politics* being shaped through these techno-environments by various actors in the city from municipal officials to social movements; and *imperatives* that come into view through a focus on these techno-environments in relation to urban futures, and the potential they may hold or need overcoming. The purpose is to generate a type of open-ended glossary of what can help respond to the key aim of the book: to account for the role of infrastructure in the third wave of urbanization with specific reference to Africa but hopefully useful to rethinking conditions elsewhere.

2

URBAN MODERNITY AND THE AFRICAN CITY

BACK TO THE FUTURE

"Modernity," as a term, has long been centered in and attributed to Europe. It has been understood as a European experience stemming from the Enlightenment, one in which, as Achille Mbembe (2017, 129), argues "the Black Man is in effect the ghost of modernity." Here, the figure of the Other is paradoxically both integral and erased. The Nigerian art critic and curator Okwui Enwezor (2010, 595) argued that modernity has two defining characteristics: "the first is the idea of its unique Europeanness, and the second is its translatability into non-European cultures." Modernity remains associated with a historical period, from the sixteenth century onward, and a series of ideas, norms, cultural practices, and representations of societies, even as its origins remain contested; Marx ([1867] 1990) finds it in the circulation of capital into modern production processes, Abu-Lughod (1989) in the pre-European global trade networks of the thirteenth and fourteenth centuries, and Gilroy (1993) in the system of slavery and the Caribbean plantation.

The dominant notion of modernity, however, presupposes toward a dichotomy between the West and its cultural, economic, and social institutions, on the one hand, and traditional societies, on the other, which need modernizing through the expansion of these systems aided by technologies and infrastructure. History from this perspective radiates only outward from the metropole. Enwezor's characterization is important for thinking through how the social sciences have been produced through both its inherent Eurocentrism and assumptions about the power to transform so-called traditional societies through technology into modern ones. It has meant the West has been placed at the center of accounts in the making of a history that are then universalized, flattening the histories of elsewheres (Bhambra 2007). One critical factor in this privileging has been how Africa has been held forth as a negative reflection of Europe and indeed modernity itself.

Why was modernity understood as emerging out of the terrain of Europe and in ways that erased Africa? We can look at historically influential ideas, such as those proposed by Georg Hegel in his *Philosophy of History* (1956), in order to see how Europe and Africa are relationally produced through the dialectic of the modern and unmodern or the complete and incomplete. Hegel argued that "Africa is no historical part of the world; it has no movement or development to exhibit." Black scholars have long challenged these Eurocentric, racist assumptions (Camara 2005; Taiwo 1998). Ronald Kuykendall (1993, 572) asked, "But why is Africa, considered by Hegel as the unhistorical, undeveloped Spirit still involved in the conditions of mere nature?" From these revised histories, European expansionism alone did not instigate globalization. Indeed, we can trace the origins of modernity in Africa back much further than the imperial enterprise (Taiwo 2010), visible in the centuries-long trading networks between the region and the world: with Asia through ocean routes incorporating more than 1,000 years of coastal Swahili culture that echo through the writing of Tanzanian novelist Abdulrazak Gurna, and toward the Islamic world through millennia-old trans-Saharan routes still traveled today by nomadic Tuareg people. Jean-Francois Bayart and Stephen Ellis (2000, 218) insist that "considered in a view of history over the *longue durée*, Africa has never ceased to exchange both ideas and goods with Europe and Asia and later with the Americas."

These precolonial histories incorporated important urban centers that acted as nodes in "the world system before European hegemony" (Abu-Lughod 1989). Take, for example, Timbuktu, founded more than a millennium ago in modern-day Mali. The city grew, as it lay at the confluence

2.1   One of Timbuktu's many libraries.

of various trans-Saharan trade routes between West and North Africa and the Arab world. In effect, the city's strategic position at the center of these networks of proto-globalization predicated its growth. Most important to this location was the Niger River, a key resource for transportation and supply of water. The wealth of the city cannot be exaggerated. Mansa Musa, the tenth ruler of the Malian Empire, made a pilgrimage to Mecca in 1324 with more than 60,000 men and tens of thousands of kilos of gold, with his estimated wealth upon his death equivalent today to USD 400 billion. We can consider Timbuktu as a "global" financial center because of its important role in the trade of salt and gold. By the fourteenth century, it was five times bigger than London and a preeminent city for Islamic teaching, and in the fifteenth century, when it was renowned for its cosmopolitanism, it had more than 200 *maktabs* (Quranic schools). Numerous libraries in the city still hold centuries-old manuscripts of this culture—a reminder of Timbuktu's time as a global city, even if its current status seems to be somewhat peripheral.

Why, then, despite these urban histories, was the civilization of Africa erased? Put simply: European imperialism. Constructions of Africa and the African as "backward," "tribal," and "savage" justified slavery, exploitation, and later colonialism by European powers. This representation was integral to the operation of the expanding world system of racial capitalism (C. Robinson 1983). We can elaborate the making of Africa as the mirror of modernity through the work of Edward Said and his exploration of the Western knowledge production of the Orient. Said (1979, 3) introduced us to the figure of the Other and the discursive representation of Orientalism in which European culture was "able to manage and even produce the Orient, politically, sociologically, militarily, ideologically, scientifically, and imaginatively." The idea of Africa as the "Dark Continent" shows how its negative reflection of Europe and modernity made it the Other, a space of incompleteness requiring technological intervention to become modern. Africa is historically frozen in a web of dualities against the "Enlightened West," through what Bayart (1993, 117) referred to as "the politics of the mirror" and Valentin-Yves Mudimbe (1988) referred to as an "alterity." Both of these terms disclose the negative categorization of Africa and the African in relation to Europe and the European.

The making of a "Dark Continent" belied the long-standing civilizations and cities of the region, such as Timbuktu. Centers of learning, trade, and political life for many centuries, these spaces were erased as part of the European project of modernity. This attempted obliteration of the urban cultures of Africa directly contradicted the accounts of the early European explorers and established the tradition of effacing the role of the region in the making of modernity. As Cheikh Anta Diop (1974) explained in *The African Origin of Civilization* through the words of Leo Frobenius, a German ethnologist who observed that "The revelations of the navigators from the fifteenth to the eighteenth centuries provide positive proof that Black Africa, which extended south of the desert zone of the Sahara, was still in full bloom, in all the splendor of harmonious, well-organized civilizations."

Western knowledge explicitly ignored the rich histories of various cities, for they challenged the underlying justification of colonialism: to bring civilization, technology, and light to a place of savagery and darkness. In the Kingdom of Kongo, established by Lukeni lua Nimi sometime after 1390, Mbanza Kongo, the capital city, was an important globalized urban space. Its urbanity and cosmopolitanism made this an important precolonial African city. Its cultures, with workshops producing pottery, metals, and textiles, were outward looking, with Kongo engaged in long-distance trade and open to the opportunities present in the Portuguese who first arrived in the late fifteenth century. The adoption of Christianity (while retaining many aspects of Kongo culture) provided a shrewd move by the Manikongo. By the early eighteenth century, Laurent de Lucques described the city (in Cuvelier and de Lucques 1953, 257) in terms that emphasized how trade and diplomatic relations had helped to forge a global city in which "The population lived in opulence because this was the metropolis of the vast kingdom where the riches of the provinces were flowing." And yet to understand the growth of the Kingdom and its city as singularly made up through this relation with a European power is to disregard the histories, social relations, and ambitions of Kongo. As Tony Green (2019, 201) noted, "Its process of growth and urbanization grew from within society not outside." Evidently, African cities such as Timbuktu and Mbanza Kongo show a "modern" history of precolonial civilization, culture, and importance in global networks of trade and exchange.

This representation of the continent would come to substantiate the notion of the unmodern city. It has found long-lasting resonance in the social sciences and ways of comprehending urban modernity, and the contemporary statues of cities in the south including, as I will explain, infrastructure. Achille Mbembe and Sarah Nuttall (2004, 348) suggested these urban histories remain "perpetually caught and imagined within a web of difference and absolute otherness." Colonial powers expunged precolonial and indigenous infrastructure and technologies acting to control nature in Africa. And contemporary urban geographies remain understood only on the terrain of the experience of modernity in the metropole, and the constructed deficiencies and absences of the African city.

Colonial expansion stepped up after the 1884–1885 Berlin conference as the Scramble for Africa drew in European powers (Youé 2015). This carving up of Africa would foremost be justified through mobilizing the Other. It was advanced through written accounts by imperialists such as Henry Morton Stanley or the geographer Halford Mackinder (1900), alongside writers including Joseph Conrad (1902) in his *Heart of Darkness*, and longer histories of representation by traders and missionaries. This Othering constructed Africa as a dark, impenetrable place, requiring science, technology, Christianity, and capitalism to become civilized. It has remained stubbornly constant in contemporary representations of the region (Jarosz 1992). The invention of Africa as unmodern, spanning centuries of racialized Othering and anti-Blackness, violently expunged the civilizations of the region. It was integral in the relational making of Europe as the modern metropole.

MODERNITY BETWEEN METROPOLE AND PERIPHERY

To break apart the assumptions inherent in Eurocentric constructions of modernity and the Othering of Africa in its making as unmodern is an important first step in thinking anew about the Infrastructural South. This postcolonial orientation demands that we ask why and with what consequences urban modernity became synonymous with the Western metropolis and understood as an unfinished technological project in those regions now termed the South. This is a dominant, problematic chronicle of urban modernity that is now universalized as a singular movement of history through modernization and the intimate connection to technology and its capacity to control the socio-natures of urban environments.

Marshall Berman (1983, 15) argued, "To be modern is to be part of a universe in which, as Marx said, 'all that is solid melts into air.'" He set out the contradictions of urban modernity present in the creative destruction unleashed in the modernization process as it transformed the Western world and ushered in the age of the modern city. Berman's (1983, 48) examination of Goethe's character Faust[1] makes explicit these contradictions at the heart of the experience of modernity: "The paradoxes go even deeper; he won't be able to create anything unless he's prepared to let everything go, to accept the fact that all that has been created up to now—and, indeed, all that he may create in the future—must be destroyed to pave way for more creation."

Urban modernity was totalizing. It required the Promethean control of urban natures and the transformations of these natures to produce the modern city. This revolution involved wholesale socio-ecological destruction, with cities such as London and Paris built again out of the ruins of the *ancien régime*: old medieval quarters, the peasantry, the blood and toil of the new proletariat, and the material waste and ruins of the earlier city. Marxist thinkers from Engels to Harvey have written about Paris as the capital of modernity, drawing attention to the violent techniques of urban planning through which Haussmann, under the auspices of Emperor Napoleon III, rebuilt the city. The contradictions inherent in the immense, violent technological and environmental metamorphosis of urban spaces are essential in the making of the modern city and reverberate out into the huge transformations currently proceeding in the South.

I have found these modern Western cities are perhaps best conveyed in all their splendor, squalor, and psychological metropolitan frenzy by novelists such as James Joyce in Dublin, Fyodor Dostoevsky in St. Petersburg, or Virginia Woolf in London. The pervasive supposed universalizing experience (at least within Western society) of being modern, and of the city, was born of the collapse of the old order. As Walter Benjamin (1997) conveyed most notably in *On Some Motifs in Baudelaire*, this collapse and subsequent technologically enabled making of the modern city brought shock and alienation as a new way of life. It is worth turning to another writer of the modern novel to establish the premise of these transformations. Boa (1996, 62), writing on Kafka's (1926) *The Castle*, described the novel as powerfully conveying the "promised progress towards ever greater control over nature and towards emancipation in a new urban, cosmopolitan culture." Arguably, then, the dynamic relations between technology and environment became constitutive of urban modernity.

THE URBAN POLITICAL ECOLOGY OF MODERNITY

Control of nature by man, enabled through technological innovations and emerging urban networks, is synonymous with the modernity of the metropole and particularly the shift to an urban way of life across the West. In effect, the deployment and operation of networked infrastructure produced urban modernity. This is despite the role of technology often being overlooked or subsumed in broader narratives on the making of modernity (Misa, Brey, and Feenberg 2003). As Maria Kaika (2005, 5) contends, the "Promethean project of modernity" was "the historical, geographical process that started with industrialization and urbanization and aimed at taming and controlling nature through technology, human labor, and capital investment." Infrastructure harnessed nature into the accumulation process and provided the techno-environmental basis of urbanization (Cronon 1991; Swyngedouw 2004). Infrastructure managed the essential resource flows required for industrial capitalism and the associated social reproduction of the proletariat. Studies have shown how these systems offered a way of controlling various deadly and circulating pathogens and disease, delivering essential urban services such as fresh water or energy, and connecting cities into networks of global commodity exchange. In effect, we can only understand the modern condition through the critical role of infrastructure in its formation (Edwards 2003)—hence the need to foreground these systems in accounts of urbanization and the retelling of how the modern was made. And to be modern meant to inhabit these infrastructures—not just in the rapidly urbanizing industrial cities of the West, but also in a myriad of ways in other places beyond the metropole.

UPE as a field of study has long provided an important tool of critical thinking about how infrastructure has enabled attempts to control nature, and its transformation into the built environment as the technological foundation of modernity. Harvey's (1996, 186) claim that "there is nothing unnatural about New York City" is perhaps the most succinct way of conveying these dynamic, techno-environmental relations that are constitutive of the urbanization process. UPE insights arguably remain central to any critical project attempting to bring into view the role of infrastructure in the making of urban modernity and the shaping of urbanization patterns.

Four propositions emerged that sustain the key concerns of this field and help shape the focus of this book on infrastructure. First, urbanization, and urban modernity, are underpinned by socio-natures in which the transformation of nature produces cities through incorporation into the capital accumulation process (Heynen, Kaika, and Swyngedouw 2006). Second, these broader transformations can be understood through the hybrid circulations of various socio-natures that center the role of infrastructure in facilitating or taming these flows (Ranganathan and Balazs 2015; Njeru 2006). Third, asymmetric power relations are always active and operate through and across infrastructure. These inequalities are staged across scale from the household or neighborhood, often described as the "everyday" (Bulkeley and Castán Broto 2013; Cornea, Véron, and Zimmer 2017; Loftus 2012) to municipal and metropolitan regions (Swyngedouw 2004) through to the planetary (Arboleda 2020). And finally, populations contest the resulting conditions; infrastructures are more than technical artifacts and are fundamental in the political life of cities and the urbanization process (Swyngedouw 2004). Urban political ecologists have put these propositions to work to interrogate the various techno-environments of urban Africa, including e-waste, construction, and alcohol through to plastic bags and water (Amuzu 2018; Myers 2008; Lawhon 2013; Smith 2001), all of which teach us a lot about how to use this important subfield of urban studies. For now, I simply wish to highlight the importance of UPE in its capacity to study infrastructure critically.

If accounts of urban modernity in the West draw on a heroic narrative of taming nature, of ordering the city, and of using infrastructure to underpin these objectives, then the postcolonial orientation set out above, combined with the insights offered by UPE, provide the apparatus to problematize and politicize these infrastructural histories, presents, and futures. To move beyond the glittering streetlights and classical architecture of the Parisian boulevard is to step into the dark side of urban modernity. This incorporates the slums razed to the ground and the expulsion of the urban poor by Haussmann's shock doctrine of capitalist urbanization. And following the dialectic of urban modernity further, it shows the modern city as always contingent on the plunder of natural resources and the exploitation of people beyond Europe (James 1938; Rodney 1972; C. Robinson 1983). To make Paris modern entailed imperial expansion and coloniality. Modernity was always Janus-faced through the system of racial capitalism and its extractions that allowed the streetlights to shine brightly into the Parisian night.

A system that swallowed up and transformed nature into capitalist value and built environment required huge parts of the planet to be brought under the system of racial capitalism through expropriation, sabotage, enslavement, and often genocidal violence (C. Robinson 1983). We acknowledge that "The enslaved African . . . is as thoroughly modern as the factory worker" (Chambers 2017, 4). Here, the notion of Europe singlehandedly instigating modernity is perverse and indeed racist, an anti-Black erasure of the pivotal role of Africans and African cities in this formation. To be clear, a critical reinterpretation of urban modernity needs to integrate better how its infrastructural making in the metropole was reliant not just on emerging networked technologies of control over nature but on the production of unmodern spaces of colonial Africa and extraction through underdevelopment and violent marginalization in the world economy (Amin 1974).

To consider AbdouMaliq Simone's (2004a) call to think about the "city-yet-to-come" leads to an important question: Is it possible to make visible an infrastructurally produced modernity across urban Africa that has a different set of coordinates than have been attributed to it—a distinct time/space through which the futures of the region's urbanization will be written? These conditions are a presage of an urban modernity whose story has yet to be fully told. The first step, as I have established, must be to recognize the inherent flaws of Eurocentric narratives in which, from the vantage point of the metropole, "the south of the world is invariably considered in terms of lacks and absences. It is not yet modern; it still has to catch up" (Chambers 2017, 61).

In African towns and cities, modernity, as understood from this Western teleological vantage point, is supposed to be an incomplete project. Edo was never modern; Accra is not yet modern; Kampala is unmodern. And the route to this urban modernity is to follow a pathway defined by the techno-environmental experience of the metropole. As J. Robinson (2006, 13) argued, "Contemporary thinking about cities silently reproduces this idea of an external 'savagery' that sustains the fantasy of (Western) urban modernity." From an infrastructural perspective, these urban spaces experience unruly natures yet to be tamed through Promethean control over flows of water, human waste, pathogens, mudslides, and so forth. Basic services remain problematic or inaccessible, pollution and contamination are central to the urban experience, and disasters and everyday toxicities threaten survival and curtail "development." Collectively, I would suggest such discourses produce a pattern of knowledge production that conveys the idea of an unfinished journey toward urban modernity. This is an incompleteness only contained in the South that is critical and central to how we imagine, plan, and understand African cities. A now well-established critique that conveys the hierarchical ranking of various cities (J. Robinson 2006) has attacked these troubling traditions in urban planning and development. And we can also think about these infrastructurally in part through the power of the networked city model.

THE NETWORKED CITY MODEL

This model, as I touched upon in the introductory chapter, has long dominated how policy makers, technical experts, and others imagine, plan, and approach the urban service provisions of cities (Coutard and Rutherford 2015; Swilling 2011; Wamuchiru 2017). As I have explained, urban networks such as the streetlights of Paris have become in many regards a form of shorthand for urban modernity writ large. We can primarily understand this model as an orderly, unitary city underpinned by safe, fully functioning, and universal networked coverage in which nature as threat (e.g., bacteria) is safely contained and nature as life-giving force (e.g., drinking water) is delivered to populations. Infrastructure is used to govern various socio-natural circulations from waste and water to electricity and sanitation through what Thomas Hughes (1983, xi) termed "coherent structures compromising of interacting, interconnected components" that transformed the built environment. As Kaika (2005, 28) argued, "Endowed with modernity's technological networks, the urban fabric became a nexus of entry-exit points for a myriad of interconnected circuits and conduits."

Literature has shown that the making of the networked city and its subsequent model arose from the heterogeneous geographies and histories of the nineteenth-century Western city. Infrastructure was subjected to logics of integration with multiple systems that previously coexisted in various ad hoc configurations that belied linear narratives of urban modernity, even in Europe (Gandy 2014). Municipalities and other actors transformed these systems into state-run, universal networks. Uniform network design, single operators providing universal services, customers that remained passive users, and oversight provided by a highly regulatory state composed this integration (Monstadt and Schramm 2017). The history of electrification in North American cities shows the multiple ways in which electricity entered everyday urban life within the context of changing technology, culture, and social relations, and as a symbol of modernity (Nye 1992). We can characterize this period by the grand ambitions of making the modern city through technology in which "systems and the men who built them became the harbingers of some naturalistic progress based on an essential and technologically determinist notion of how infrastructures related to cities" (Graham and Marvin 2001, 47). Despite work showing how these systems have "splintered" under neoliberal economic conditions in the West, the model remains dominant in investment and engineering plans and a staple of linear histories of technology, and the experience of urban modernity is complete.

Returning again to the work of Jennifer Robinson (2002, 2), I want to emphasize how this kind of discursive representation acts "to ascribe innovation and dynamism, modernity to cities in rich countries, while imposing a catch-up fiction of modernization on the poorest." Over recent decades, scholars, including many of my colleagues, have argued that the networked city model is flawed based on the reasons set out above. But such critique requires more than problematization; rather, I would affirm a retelling of the histories, presents, and futures of urban infrastructure in the region. Critical to this task is to make clear the extraordinary power of the model in the governing of urban networks amid the diverse, ever-shifting geographies of the Infrastructural South. Echoing Enwezor's (2010) characterization of modernity through its translatability from Europe, Jochen Monstadt and Sophie Schramm (2017) termed this process the translation of the model by planners and engineers across various eras of urban governance. This incorporates thinking about transferring the model to colonial urban space and its exclusionary premise, in which service provision was racially segmented.

Baganda, the dominant kingdom in modern-day Uganda and one of the most powerful in Central Africa by the end of the nineteenth century, provides an example of the effacing of precolonial planning through colonial planning logics incorporating a particular vision of the model. British colonial officials understood Baganda, even if they did not publicly acknowledge such as having a highly organized state, supported by an efficient bureaucracy (Nziza et al. 2011). The seat of power was a well-ordered, mobile settlement of 40,000 until the reign of Kabaka Mwanga II, who established on Mengo Hill the grand palace of Kibuga surrounded by kilometers of walls and connected to the hinterlands by an intricate system of roads and transportation routes on Lake Victoria. Indeed, "the organisation of the indigenous capital represented the delicate and highly structured order of Ganda political society" (Monteith 2019, 251).

The sophisticated town planning of precolonial Kampala counters the colonial justifications by Britain of bringing civilization through technology to Africa. And if we consider how Baganda managed its resource flows, there is evidence of an efficient sanitation system. As Stephanie Brown (2014, 18) has written, "the Baganda were orderly and sophisticated in their toileting habits." British imperialists such as Samuel White Baker may have wondered whether the city was more hygienic, safe, and provisioned than the slums of "modern" Victorian England. Colonial authorities ignored these infrastructural conditions in accounts that sought to justify the need for Enlightenment principles seemingly (and paradoxically) embodied in colonial subjugation and deployment of new technologies to control urban environments. In colonial Kampala, the planning of sanitation and water services translated the network city model but was only concerned with its delivery in the spaces of the British colonizer. Mirams' Kampala Plan of 1930 later reflected this disposition, which was perpetuated in the *Kampala Plan of 1951*, neglecting the infrastructural needs of the colonized in ways that shaped a "dualized city" and "reinforced the topographical expression of colonial power" (S. Brown 2014, 79).

Colonial rule shaped the planning discourses and logics that structured how rulers built and administered colonial cities, such as Kampala (King 1977). This mirrored the broader, constructed relations of modernity between a "backward" Africa and an "enlightened" Europe. Legislation, segregation, and planning all reflected a hierarchy of control across these colonized urban spaces, predicated on widespread subjugation of African urban populations (F. Demissie 2007; Yeoh 2001). This is a history of the networked city model mutated through the logics of the white supremacism of colonial governance. We can understand this in two ways that convey the experience of urban modernity in the south. The first is through ignoring African populations and the infrastructural needs of these areas (Myers 2006). As the colonial governor of the Gold Coast in 1858 stated, "the object of this Government was not to clean out dirty towns but to direct the people to that and other objects by controlling and modifying their own Government" (in Hess 2000, 40). Second, other forms of urban colonial governance, particularly as the colonizer expanded his presence in these cities, displayed a more active logic of control through the planning and operation of infrastructure, especially focused on health ordinances, fears of contamination, and prevalence of pandemics to segregate populations (Swanson 1977). For instance, the *1878 Gold Coast Towns Police and Health Ordinance* created a new legislative tool for the authorities to empower the governor to deal with unsanitary conditions and impose punishments that included fines and court appearances upon those deemed to be contributing to unsanitary conditions.

Frantz Fanon (1967) gives us a sharp, critical analysis of the spatial and infrastructural separation that underpinned colonial urban space in *The*

Wretched of the Earth. Fanon's description of the division between the "settler town" and the "native town" in these colonized spaces makes clear the racist planning logics and only partial implementation of the model through which African cities were governed. He wrote, "The zone where the natives live is not complementary to the zone inhabited by the settlers. The two zones are opposed, but not in the service of a higher unity . . . they both follow the principle of reciprocal exclusivity" (1967, 39).

Fanon described the material differences between the two zones. He understood the settler space as "a strongly built town, all made of stone and steel . . . a well-fed town, an easygoing town; its belly is always full of good things." This was in contrast to the space of the colonized, "a world without spaciousness . . . a hungry town starved of bread, of meat, of shoes, of coal, of light" !967, 38) Fanon's work is important not just as a historical record for unequal cities such as Accra or Cape Town, but also in pushing scholars to consider how these colonial spatial legacies continue to mutate into the everyday infrastructural conditions of the present and the lives and experiences of the urban majority. This is a present in which some remain without sufficient food, access to resource flows, or technologies. We can trace and make visible these processes into the contemporary era. This is because infrastructure continues to act as the critical materiality and technology for controlling urban populations for the benefit of the ruling elite. Infrastructure is the materiality of inequality across the techno-environments of African cities. To undertake this tracing means to integrate the "imperial durabilities of our times" (Stoler 2016), demanding we think about how contemporary conditions remain steeped in and spin out of the colonial histories of implementing the networked city model but not entirely defined by such processes.

The colonial planning logics that produced segregated infrastructure shaped a legacy of highly fragmented, racialized cityscapes across the South, in which the colonizers did not complete the plans set out in the networked city model (Kooy and Bakker 2008). Some studies have highlighted these ongoing spatial legacies and afterlives of colonial-era infrastructure and planning, in which new rulers continued the model into the modernization programs of the postindependence era. The persistence of the highly segregated premise of the model in these contexts is critical for understanding how inequalities in distribution, provision, cost, safety, and experience

2.2 Cape Town, the "settler town."

2.3 Cape Town, the "native town."

continue to shape infrastructure in the contemporary era. For instance, Wangui Kimari (2021, 1) brings these histories to life through the story of a water pump in Mathare, Nairobi, to show how this infrastructural artifact is more than a piece of hardware but rather an "articulation of an enduring colonial political economy in Kenya."

POSTCOLONIAL MODERNIZATION

In foregrounding the histories of infrastructure in relation to the networked city model, it is also important to revisit the postindependence period of modernization. The end of colonial rule shifted the basis of infrastructure from its highly segregated premise in cities and its extractive logic beyond the city toward a new development trajectory for African nation-states. In many regards, it seems these systems remained consistent with the colonial enterprise of civilizing and making modern the people of Africa through and with technology (Kimari and Ernstson 2020).

Ngũgĩ wa Thiong'o's novel *Petals of Blood* (1977) evokes the power of infrastructure in the making of modernity in this era. The reader learns about the lives of several protagonists—Munira, Abdulla, Wanja, and Karega—as they retreat to the village in the shadow of the Mau Mau rebellion and attempt to deal with a new, rapidly modernizing, and independent Kenya. Thiong'o portrays modernity most prominently through the deployment of a road as the postcolonial state builds a "Trans-Africa" infrastructure through the village. The construction of the road sets in motion the contradictions and paradoxes of the modernization experience and the "shock cities" of the era that later find resonance in the work of Berman. Thiong'o (1977, 312) wrote, "The road brought only the unity of earth's surface, every corner of the continent was in easy reach of international capitalist robbery and exploitation."

With such modernization and integration into the world economy comes creative destruction. As Berman would say "all that was solid melts into air." Thiong'o (1977, 335) wrote, "Occasionally, the Town Council has a clean-up, burn down campaign. But surprisingly it is the shanties put up by the unemployed and the rural migrant poor which get razed to the ground." And we see this in the tragic endings of Munira, Abdulla, Wanja, and Karega as Kenya integrates a post-independent modernity and associated capitalism at the expense of tradition and as the city begins "to encroach upon and finally swallow the traditional and the rural," as Williams (1999, 78) has argued. It is the infrastructural artifact of the road that becomes the key protagonist in this march of modernity.

In the city, despite promises of delivering universal infrastructure services for expanding urban populations through incorporating the logics of the networked city model, the techno-environmental inequalities of the colonial era were in part sustained. Kefa Otiso and George Owusu (2008, 150) analyzed this history in urban Ghana and Kenya. They argued, "The provision of housing, basic services, and urban infrastructure also suffered in both countries because of their continued reliance on colonial urban planning regulations, by-laws, architectural styles, and housing standards." The replacement of the colonial elite with a new postcolonial ruling class meant in many regards the maintaining of the status quo (Konadu-Agyemang 2001). Colonial-era laws, regulations, and ordinances, rather than being replaced, became an opportunity to take control of former colonial residential areas, accompanying infrastructure services, and the privileged lives of the colonizer. Later in this book, I will return to chart further how these infrastructural histories produce contemporary conditions. For now, the point I wish to make is that infrastructure was both symbolic of postindependence modernity and indicative of its material technology, and an attempt at control over nature and people as they become collateral damage in the making of new urban worlds.

I have briefly set out the unequal, spatially differentiated histories of infrastructure in urban Africa in relation to the networked city model and across the colonial and postcolonial eras. I have argued the ideas contained within the model have continued to shape how we understand and govern urban networks into the contemporary moment and within the broader coordinates of the Infrastructural South. AbdouMaliq Simone and Edgar Pieterse (2018, 29) argued, in thinking through the music video *Capture* by Congolese-Belgian musician, Baloji, for the need for researchers to make visible and incorporate the "co-existence of a colonial, neo-colonial and bastardized present, reaching forward to circumscribe the future." And this must include how understandings of infrastructure within the variegated experiences of urban modernity have themselves incorporated such coexistences.

The networked city model has become the most prominent vision and aspiration for planners, engineers, and politicians in the planning of urban-scale infrastructure. It is a marker of progress and developmentalism, and it is, above all, a device through which donors, governments and policy makers understand whether cities are modern. The model, then, is a powerful discourse through which cities have sought to mobilize resources in order to move toward a universal, homogenous system that has become a dominant objective of urban planning for infrastructure in the South (Nilsson 2016). In this linear narrative, noncentralized networked infrastructure (whether "formal" or "informal") is merely a premodern stage in development toward the techno-uniformity and urban modernity that the model promises (Smiley 2020). As a shifting, material geography, these technologies have therefore helped decipher the relative status of particular cities in the broader categorization and labeling of different urban spaces (J. Robinson 2006). Urban-scale infrastructure has acted as a supposedly easy-to-discern materiality of progress and modernity, a crucial barometer of development for the global institutions of governance and finance such as the UN-Habitat and the World Bank.

Several studies, including work I have undertaken collaboratively, have focused on actual existing conditions beyond the model and questioned whether it provides the most productive way to think through the Infrastructural South. This has proposed terms including "heterogeneous" (Munro 2020; Rusca and Schwartz 2018), "hybrid" (Furlong 2014; Jaglin 2019), and "peopled" (Doherty 2017; Simone 2004b) to describe these conditions and built on the premise "to see infrastructures as emergent, shifting and thus incomplete" (Guma 2020, 728). This incompleteness proposed by Prince Guma connects to the broader implication of urban modernity as unfinished, as well as making visible the ever-shifting "always in the making" (Baptista 2019) dynamics of urban networks.

Scholars have also suggested that the model has a detrimental effect in developing alternative infrastructures for delivering essential urban services by framing these modes of provision as nonmodern (Nilsson 2016). This is an important implication because it shows that concepts and ideas concerning infrastructure have a real-world effect (Castán Broto and Sudhira 2019). Until we disassemble dominant traditions of knowledge production and

2.4 A toilet in Namuwongo, Kampala: one particular heterogenous infrastructure configuration.

propose new models, planners, policy makers, and politicians will struggle to think anew about techno-environmental futures for urban Africa (Pieterse 2013). This supposedly unfinished project of urban modernity (in the periphery but not metropole) and the focus on attaining the features of the modern networked city mask the actual heterogenous, fragmented conditions and operations of infrastructure (Furlong and Kooy 2017).

However, many of the most popular scholarly, policy, and engineering approaches to infrastructure remain welded to this networked city model and the associated logic of a linear trajectory toward a universal, modern, and completed system at some future point. This is despite the existing geography of partial, hybrid infrastructure systems throughout cities of the South (Furlong 2014; Guma, Monstadt, and Schramm 2019) as well as the ever-shifting configurations of infrastructure also in operation in the North. As I made clear in the previous chapter, the lives of tens of millions of urban dwellers are intimately connected to the planning and

management of existing and future planning of infrastructure. The conno-
tations for how various actors think about infrastructure are considerable
for our global future.

THE INFRASTRUCTURAL SOUTH

COUNTERHEGEMONIC NOTIONS OF MODERNITY

How might we think anew about urban modernity beyond these exhausted,
discredited tropes in a way that not only overcomes a Eurocentric narrative
of infrastructural history, but also articulates the contemporary topographies of Africa's urbanization? Or as Iain Chambers (2017, 107) questioned,
"Where and how are we to locate these unrecognized epistemologies of
modernity?" Rethinking the Eurocentric conception of infrastructure, and
these problematic traditions of Othering Africa as the unmodern, is a necessary step in setting out the concept of the Infrastructural South. I draw
from and advance two intellectual traditions that have problematized the
Eurocentrism of modernity to advance an alternative conception. We can
broadly characterize these as "counterhegemonic notions of modernity"
(Ejigu 2014). Various interpretations begin from the straightforward premise that we can no longer sustain the fallacy of a singular experience of
modernity amid the diverse histories of its unfolding. Modernity is not the
result of transformations contained solely within the West; instead, modernity is imbued with the ever-shifting relations, conflicts, and flows of the
modern world in its entirety.

Work that has extended these ideas included the drawing of attention
to "alterative modernities" (Ashcroft 2009) and "multiple modernities"
(Eisenstadt 2017). Collectively, if there are differences, these ideas all seek
to emphasize the fallacy that modernity was a Western-only phenomenon,
or indeed that there was even a singular modernity operating in the diverse
sociopolitical landscapes of Europe. We can bring these ideas forward to
show the importance of considering the grounded, localized, and variegated ways in which modernity unfolds across time and space and exists
in a multiplicity of hybrid forms. This is a dialectic between the global and
the local, the modern and traditional, and the complete and incomplete.
As Shmuel Eisenstadt (2000, 2) asserts, "Such patterns were distinctively
modern, though greatly influenced by specific cultural premises, traditions,

and historical experiences." To be clear, I am arguing there was never an originary urban modernity; rather, there is only an urban modernity that has been bound up with the contradictions, negotiations, and mutations in its encounter with particular spaces and histories, both inside and outside of Europe.

Debates on Afro-modernity help to advance these counterhegemonic ideas of modernity with specific reference to the region in thinking about contemporary urbanization. They seek to explain the global experience of modernity as a complex, hybrid interplay and negotiation. Sembene Ousmane (1995) sets his novel *God's Bits of Wood* in 1948 French Colonial West Africa and follows a strike on the Dakar–Niger line. Melissa Myambo (2014, 465) argued that the novel showed how workers "enact a complex relationship with the technology of the train, the 'machine.' While the railroad symbolizes colonial capitalism, imperial modernity and racial oppression, it also symbolizes liberation from all of the above. . . . The machine is here and, therefore, theirs to appropriate as it suits them because of, and in spite of, its origins." In this passage, we see the ways in which modernity was constantly being worked out and negotiated with complex intersections that brought both trauma and opportunity, resistance and appropriation.

In this view, the repurposing of a technology of modernity illustrates "Africans ability to negotiate modernity on their own terms" to "produce their own social reality in dialogue with modernity as they move from colonialism into a world defined by themselves, and by what they do in their everyday life" (Macamo 2005, 4). These always-shifting, messy encounters question whether Africans and Africa can ever be modern when we appraise them against the supposed (but false) universalism and completeness of the European experience. Concurrently, these accounts question whether analytical frameworks devised from this Eurocentrism are even adequate for understanding the diversity of the modern world and the ways in which modernity was constantly being worked out in many often contradictory and constrasting ways across different spaces. Francis Nyamnjoh (2001, 364), in his reading of James Ferguson's (1999) *Expectations of Modernity*, offers important guidance by suggesting, "The way forward is through a re-reading of a misrepresented past and present." To do so, I argue, means to engage modernity differently, to situate ourselves in the "postcolonial present" (Chambers 2017, 3) in ways that destabilize, break apart, and disrupt the hegemonic traditions of knowledge production, including across urban studies.

Collectively, the broader literature on multiple or alternative modernities, and the specific work focused on rethinking modernity in the African context, opens up a new way to interpret the Infrastructural South as a location or condition and to mobilize it as an epistemology. We can center this on the rejection of Eurocentric understandings of modernity while recognizing the power of the idea of modernity—civilization and progress enabled through technology in shaping the making of Africa—and focusing on how this making has proceeded in constant negotiation with the diversity of cultures, traditions, and ways of knowing and living. This mode of thinking, the broader experience of modernity, becomes an anchor point for disrupting the dominant narrative of infrastructurally produced urban modernity in which a linear, teleological, and Eurocentric way of knowing should no longer hold sway.

My response to problematic traditions of understanding urban modernity, and the necessity of locating an epistemological way forward, emerges from counterhegemonic notions of modernity, alongside a broadly postcolonial orientation and critical analysis rooted in the techno-environmental concerns of UPE. I advance this work specifically through the concept or framing of the Infrastructural South.

The Infrastructural South conveys the myriad, ever-shifting relations constitutive of urban modernity. In doing so, I bring infrastructure into view as the means to tame, formalize, and accumulate from socio-natural circulations across urban space. These systems are therefore a critical factor in how broader urbanization trajectories are and have been shaped. Ravi Sinha (2012, 4) argued in specific reference to a mode of thinking through socialist futures that "Despite famous theories of convergence, which expected all societies of the world to become progressively like those in the west, the spread of modernity, even after the onset of so-called globalization, has not been able to homogenize the world."

The Infrastructural South denotes an alternative experience and future far removed from what Ferguson (1999, 14) critiqued as the (Western-centric) "teleological narratives of modernity" that became wrapped up in the unfolding modernization process in post-independent nation-states such as Zambia. There is, in effect, no linear, fixed pathway toward the homogenized, completed future. Ferguson (1999, 5) argued that "Urban-ization, then, seemed to be a teleological process, a movement toward a known end point that would be nothing less than a Western-style indus-trial modernity." We know this to be a fallacy—that the urbanization process, like urban modernity, requires a conceptualization that can grap-ple with abstract/grounded and global/local tensions in the histories and geographies that lie beyond the Eurocentric narrative. To do so brings into view the multiplicity of ways in which people and places have negotiated, navigated, and refashioned these transformations (Mitchell 2000). This was and indeed still is urban space as a patchwork, hybrid assembling of modernity in which "There were places to 'keep tradition alive' and there were places to be 'modern,' places to be a 'kinsman' and places to be a cos-mopolitan urban 'dweller'" (Simone 2001, 21).

The Infrastructural South extends our analytical apparatus further into these spaces to explore how the urban modernity experienced in the West through a dialectical relation with the Other finds a different resonance in those regions we term "the South." It is a provocation to think anew about the role of infrastructure in the histories, presents, and futures of urbanization beyond the West and, as I will speculate later in the book, also on the terrains of the metropole itself.

Work by Ravi Sundaram (2009, 2) on "pirate modernity" offers one way of demonstrating this approach. Sundaram highlighted how the Southern city is a "technologized urban experience" vastly differentiated from the supposed Eurocentric markers of modernity and yet is simultaneously also modern. This is the urban condition in which "Most city dwellers in India have grown up with the rhythm of technological irregularity, the ingenious search for solutions." It is an experience of urban modernity, but one in which people are always getting by, waiting for the next interruption, nav-igating, negotiating, and waiting upon the promises of a well-run future amid what Sundaram termed "the fragmentary time of infrastructure." In essence, "Pirate culture has utilized the technological infrastructures of the post-colonial city: electricity, squatter settlements, roads, media networks and factories. These are siphoned off or accessed through informal arrange-ments outside the existing legal structures of the city" (Sundaram 2009, 12).

This pirate culture is at play in the work *ARTicle14* by Beninese artist Romuald Hazoumè, who repurposes a trader's cart, replacing the normal goods of drinks, cigarettes, and so forth to be found on the streets of Porto-Novo with the technological debris discarded by the West. *ARTicle14* refers to the urban myth or rumor of a final "fourteenth article" in African constitutions, "Débrouille-toi, toi-même," which translates as "when all else fails, do what you need to do, because no one else is looking out for you." Hazoumè's work evokes the pirate cultures and arguably the different experiences of urban Africa. In particular, the work emphasizes urban dwellers' transformation of material cultures through resourcefulness, the constant appropriation and reengineering of Western "modern" technologies, much like the rail workers in Ousmane's novel, and the need for many to hustle to get by and get on amid the ever-shifting topographies of the urbanization process.

The pirate modernity argument provides an important reinterpretation of the urban worlds of the Infrastructural South but, I would argue, needs to be extended to think more closely about the dynamic relations between technology and environment across urban space, as Hazoumè's art or Ousmane's novel alludes to. This is because everything in the world is arguably already modern and unfinished, including these "pirate cultures" and incomplete infrastructure projects that are always in a state of flux, becoming, and mutation. Africans navigate urban modernity through and by technology, transforming its intended form through their everyday interactions with the infrastructure of the city.

A NEW SET OF COORDINATES

Let us take the next step together in thinking anew about the Infrastructural South. This involves thinking through an apparatus. We need to get a grip on how these techno-environments are mutating across the broader geographies of modernity and urbanization. I have argued that dominant modes of thinking about infrastructure find their limits in attempting to think beyond the terrain of the abstracted Western city. This is part of what Pieterse (2010, 218) termed the "epistemic challenge" of addressing "the specificities of African urbanization." Therefore, chapters 3–9 in this book respond to the urgency to think through, decipher, and account for actually existing conditions of infrastructure across the urbanization process.

I have written this book with the purpose of advancing the multiple techno-environments in the making of the Infrastructural South. In each of the chapters, I focus attention on one particular techno-environment. I do so with the intention of shifting beyond a Eurocentric tradition in urban studies. I make no totalizing claims about this mode of analysis or its capacity to unpack the dense, complex, and sometimes contradictory urbanization of Africa. Rather, I engage with the infrastructure that has come into view through more than a decade of learning and unlearning focused on these, at times, mundane and, at other moments, dramatic urban networks. These techno-environments collectively understood through the Infrastructural South offer a provisional set of coordinates from which to think anew about some of the most pressing global issues of our times.

3

INFRASTRUCTURAL SECURITY TO ECO-SEGREGATION

SEPARATED CITIES

Cité du Fleuve should grow to a population of 250,000 residents. Built through land reclamation of marshes and sandbanks along the Congo River, it lays in immediate proximity to but separate from the megacity of Kinshasa. Cité du Fleuve is designed as a mirror of the unmodern city conjured out of the powerful waterway (De Boeck 2011). According to the developer, the scheme "began as a dream" and is now being materially assembled, with USD 100 million invested into land, construction, and a dense network of buildings. Its spatial form is best understood as an enclave of high-tech infrastructure that offers security, reliable services, and a life separated by water from the trials and tribulations of the adjacent city of 10 million Kinois. The making of this enclave raises a critical question about the future of Kinshasa: How will the desire for infrastructural security shift the underlying socio-spatial composition of the city? Its creation from reclaimed land provides a powerful symbol of both its presence in and its absence from the city. As Filip De Boeck (2011, 277) argued, Cité du Fleuve "replicates the segregationist model of Ville and Cité that proved so highly effective during the Belgian colonial period" while operating in a very different context of rapid urbanization, flows of real-estate finance into the built environment, and postcolonial urban governance.

Another new urban development separated by water from a congested megacity is Eko Atlantic, next to Lagos. Eko Atlantic is part of a series of massive investments in infrastructure projects across the Lagos region, incorporating the Lekki Free Trade Zone and Expressway, alongside the Lagos Rail Mass Transit system (Sawyer 2019). Assembled from 10 million m^2 of reclaimed land on Victoria Island and protected by an 8.5-km seawall from the storm surges of the Gulf of Guinea, the city is an example of the scale of transformation taking place. Central to the sales pitch of its developers, South Energy Nigeria Limited is set to be a "city within a city" through the infrastructural promise of a "self-sufficient and sustainable" space that "includes state-of-the-art urban design, its own power generation, clean water, advanced telecommunications, spacious roads" (Eko Atlantic, n.d.). Several Nigerian and international banks, including FCMB, First Bank, Access Bank, BNP Paribas Fortis, and KBC, financed this enclave, which the developers declared will house 250,000 residents. Concerns about Eko Atlantic's socio-ecological impact on the surrounding region include how it will reshape "the physical, economic, political, and socio-cultural 'risk-scape' of Lagos" (Ajibade 2017, 85), as it compounds coastal erosion and storm surges for existing mainland communities. Its infrastructural security is accessible to those who can afford it and comes with costs for those who cannot. Critics often cite Eko Atlantic as an example of a forthcoming eco-segregation that will splinter and separate space across cities as the climate crisis intensifies (Goodell 2018; Mendelsohn 2018; Tuana 2019), forming divisions across populations. Cité du Fleuve and Eko Atlantic are but two early examples of transformations in urban form that are producing *enclave techno-environments* through new city planning and large urban development sites.

The informal settlement has become the ubiquitous if narrow proposition to convey the dominant spatiality of an in-the-making urban Africa, predicating problematic representations of the region as a vast, unending landscape of poverty, slum, and technological deficit. This representation serves to tie African cities to ideas of being unmodern. As Pieterse (2010, 207) argued, models in urban studies and planning posit that "urbanism is largely equated with complex, social, natural and material interactions that unfold in western cities, whereas non-western cities are only good for describing absences and wanting." Urban studies literature and global urban policy define African cities in relation to the experience of modernity in the West. In doing so, this erases how techno-environments unfold beyond a teleological, linear narrative of history that radiates outward from the metropole.

These representations of the techno-environments at play in the region are rooted in assumptions that urban modernity remains incomplete. They mask the sweeping transformations that are proceeding through large-scale urban development projects. Design of these spaces share more in common with schemes in Dubai or Singapore than in neighboring districts within the same cities or in historical experiences of urban change in the North.

In recent decades, planning and construction of suburban compounds through to entire new cities has become visible in some parts of urban Africa. Developers promote these master-planned developments as privatized solutions for urban regions facing increasing environmental crises, population growth, and, for many, substandard urban service provisions. Promotional images of these schemes convey the allure of a pixilated, rendered urban modernity for potential customers. These promises of modernity are being assembled as a response to perceived technological dysfunction and deficit. As the much-missed Vanesa Watson (2014, 216) argued, new cities are "underpinned by the ideal that through these cities Africa can be 'modernized.'" In effect, this model, drawn not from the West but from those regions termed "the South," is being deployed as a shortcut for African cities to attain conditions of urban modernity, specifically through integration of innovative technologies into everyday life.

A topography of the urban enclave shapes the construction of these spaces. Built on the assurance of controlled environments (Marvin and Rutherford 2018), these urban spaces generate a series of operations across infrastructural life that conquer the techno-environmental challenges associated with a not-yet-complete modernity. As Ayona Datta (2016, 3) clarified, "these new cities aim to bypass the failures of existing megacities." The term "enclave" emphasizes the technological configurations of newly built environments, often with helpful tax arrangements and restricted entry, alongside separated, decentralized, high-tech, and often sustainable urban systems, and often built separated from the traditional, incomplete, and networked geographies of the extant city, reducing dependence for residents as they opt into privatized archipelagos (Graham and Marvin 2001). The enclaves being envisaged propose a radically different techno-environmental ordering and operation of the region's infrastructure from what has come before.

DRIVERS OF NEW ENCLAVES IN AFRICA

Enclave techno-environments have a long, variegated global history stemming from the colonial planning of urban spaces (Njoh 2008). In recent decades, the enclave became associated with suburbanization and securitization as elite/upper-middle-class communities opted out of the city (Marcuse 1997). We can closely tie the growth of enclaves in contemporary urban Africa both to these histories and to global, urban policy mobilities of new cities. Circulation of sustainable urban development models intersect with attempts at the "worlding" of cities in the African region (Roy and Ong 2011). A range of interests, from governments and municipalities to financiers and developers, are keen to project images of modernity and progress. Developers have embraced the influence and tropes of international design, and especially the master plans of new Asian urban enclaves (Pow 2011; Datta 2016) as a pathway to achieving such modern desires. The success in the uptake of the enclave model as a new city lies partly in how it acts as a mobile construct that crosses, touches down, and transforms in various spaces, primarily across the urban worlds of the South. As Femke Van Noorloos and Maggi Leung (2016, 127) argued, "rather than western ideas and finance being exported across the globe, Asian models, concepts and investments are influencing the remaking of today's global South cities."

Master-planning teams and international design consortiums have facilitated the circulation of Middle Eastern and Asian models of the enclave, which have been integral to the emergent technological and environmental characteristics of these spaces. For instance, Rendeavour, the developer of many large-scale developments in Africa, has used Skidmore, Owings & Merrill (SOM) LLP, an enormous global architecture, design, and engineering company, to lead projects in Kenya and Nigeria. The firm had previously designed new city projects, including King Abdullah Economic City, outside Mecca, and Baietan, Guangzhou, highlighting how plans first made for a Chinese or Saudi city may reappear in mutated form as a new Kenyan city.

If these new enclave city models are in some ways abstract, land-anywhere, and placeless typologies, it is important to highlight two points in relation to the broader argument of the book. First, enclave urbanism does not originate from the technological terrain and planning models of Western cities but rather reflects the aspirational typologies of Asian cities. Aihwa Ong (2011, 2) argued, "Today, Asian cities are fertile sites, not for following established pathways or master blueprint but for a plethora of situated experiments that reinvent what urban norms can count as 'global.'" To clarify, I suggest that South-to-South transfers of finance, knowledge, and models of urbanity are crucial to configuring enclaves in urban Africa. The West is no longer the blueprint or future destination for African cities.

Second, whether these new urban spaces are financed from Singapore, designed in South Africa, or built by a Chinese construction company, they become integrated into urban Africa: existing real-estate markets, consumer tastes, market demands, land dynamics, and the broader political-ecological histories of particular cities. Enclave urbanism as an imported model will also have distinctive African forms reflecting how the experience of modernity in African cities has long been one of negotiation and navigation, selective incorporation, and local reinterpretation. The enclave may be a pixilated, imported, and abstract model of modernity, but it becomes variegated in its encounter with the time and space of urban Africa.

We can attribute the integration of the enclave into spatial imaginaries through how the enclosure of urban land is speeding up (Gillespie 2020). The secondary circuit of capital (Harvey 1978) has become entangled in a global network of land deals and surging financial investments (Obeng-Odoom 2015b; Goodfellow 2017). Circulations of finance, generated by the elite from local resource extractions, alongside waves of rent-seeking finance from new centers of global power in the South are the primary drivers of this transformation. Drawing on what David Harvey (2001) has termed the spatial fix, that is, how capital seeks to invest in spaces of profitability, certain African cities now become sites for rent-seeking finance, which exert a powerful influence on development outcomes. If extraction from Africa has historically centered on natural resources, in the twenty-first century, this is now being joined by the massive land deals and urban development found in cities from Addis Ababa to Yaoundé. By 2015, the international real-estate consultancy Knight Frank (2015) was identifying Africa as "the frontier of growth" for global property investment.

Tom Goodfellow (2017, 789) argued that speculative investments into urban land have become more attractive due to market conditions in African countries in which "the difficulty of making the primary sector profitable . . . inflation rates tend to be high, while stock and bond investment options are limited or non-existent." This financing is critical to the creation of enclave techno-environments. Out of the twenty-six new cities I identified as in progress, at least twelve show evidence of financial involvement beyond Africa. This accounts for nearly 50 percent, and this figure is likely higher, given the opaque nature of global real-estate finance. The enclave may be the foremost material and spatial expression of the boosterist projections of an "Africa Rising" emerging from the turn of the Millennium (Mahajan 2011). The *Financial Times* attributed this rise to a combination of economic growth, increasing incomes, and importantly, the consolidation of the middle class. Franklin Obeng-Odoom (2015a, 234) offers a critical counterpoint, arguing that these narratives are instead about "confirming that Africa is ripe and ready to host investment and to open up markets in areas where they did not exist or existed but were not capitalist in form."

With real estate functioning as a crucial investment space for global capital, the opportunity presented by Africa's rapid urbanization has become increasingly aligned with the trillions of dollars of rent-seeking, finance-identifying opportunities and new so-called frontier markets (Sassen 2014). As examples illustrate, this market incorporates a myriad of actors, including private equity firms, Chinese state and non-state finance, high-wealth individuals, national and international banks, and various Southern-based investment and sovereign wealth funds. Such urban processes share some commonalities with the Asian experience that has proceeded over the last three decades and the speculative urbanism even though they remain specific to the urban contexts in which they are situated.

Third, evidence shows an increasing demand for elite and middle-class housing. This, I suggest, is intimately connected to the surging wealth of certain parts of society that allow for property investment (Mercer 2017).

Developers have shaped marketing strategies around a global modern life-style aesthetic. In Ghana, the elite (who were determined through their capacity to spend more than USD 20 per day by the African Development Bank in 2011) make up around 450,000 of the country's 30 million population. These households are shaping new consumption patterns that are reinforced in the housing sector. The variety of housing that is being constructed in certain cities, such as Accra, reflects growing spending capacity. These projects range from detached housing costs that are upward of USD 450,000 for three-bedroom houses in desirable neighborhoods, such as Cantonments, to new apartment blocks in Airport City, with prices starting at USD 250,000.

Crucially, the burgeoning middle-class African elite have joined the elite. A study by Abebe Shimeles and Mthuli Ncube (2015, 189) that assessed thirty-seven African countries determined that most "experienced a rise in the size of the middle class over the last decade." The term "middle class" remains a subject of fierce debate (see Mercer 2014; Ncube and Lufumpa 2014), as the group's income range is understood as a daily per capita consumption of USD 4–USD 20. The upper middle class may more accurately represent the influence of growing wealth in African cities. The African Development Bank (2011) suggests that this group has a daily per capita consumption of USD 10–USD 20 and makes up 13 percent of the population in Ghana. These Ghanaians have increased their spending on consumption, such as cars, household goods, flights, and housing. Together, the elite and the upper middle class of Ghanaian society, besides countries with similar levels of economic growth, are fuelling the demand for new housing even with less than one percent of the population using mortgage based finance (Gyamfi-Yeboah, 2021). It is these transformations that are essential in the development of the spatial "product" of the enclave.

Finally, we can understand these new spaces as a response to the perceived dysfunction of African cities. The enclave is a promise of infrastructural security and on-demand services for households living in cities where disruption to electricity or the water supply remain an everyday concern. This infrastructural fix has arguably come to supersede concerns about security for purchasers. Studies explained earlier eras of enclave-type development as gated communities as developers sold the housing on the promise of security for residents in cities with high levels of crime or political violence (Asiedu and Arku 2009). However, Obeng-Odoom's (2014, 551) research in Ghana and Malaysia countered this understanding by arguing that "while security is an important reason, it is the provision of quality housing services that is reported as the single most important reason for living behind gates." Infrastructure is clearly critical to the "quality housing services" that Obeng-Odoom attributes as the primary reason that residents choose to move to these spaces.

Mike Hodson and Simon Marvin (2009, 194) proposed the term "urban ecological security" to denote how "attempts to safeguard flows of resources, infrastructure and services at the national scale" have become increasingly identified at the urban scale. They argue that "increasing concerns over 'urban ecological security' are now informing strategies to reconfigure cities and their infrastructures in ways that help to secure their ecological and material reproduction." This infrastructural dynamic draws attention to how cities attempt to govern, control, and secure the flow of resources beyond the national scale to address the untamed natures that produce urban crises (e.g., intense flooding or the scarce supply of electricity). Enclaves and new cities or extensions push us to think about how these spatial endeavors are currently proceeding at both the urban scale and, importantly, within the heterogeneous geographies of cities themselves.

Infrastructural securitization is driven by private investment through decentralized, off-grid, and pay-for-access systems. In order to differentiate this process from broader and frequently state-led attempts at the metropolitan scale, one must look at the concept of "urban ecological security" and understand it as a process also occurring within the city itself. Promises of infrastructural security and the allure of managed, controlled environments are arguably integral to the growing demand for enclaved life. We can understand the premise behind these privately run, annexed enclaves as attempts to produce accumulation opportunities in addressing the challenges related to the infrastructural deficit. By starting anew and beyond the congested, poor, and politically complex geographies of existing cities, the enclave becomes a start-again fantasy of city making, with the capacity to sell its infrastructural security to those that can afford to access premium services.

THE GEOGRAPHY OF ENCLAVES IN AFRICA

Vanesa Watson (2014) understood contemporary enclaves and new cities as an African urban fantasy—an aspirational imaginary of a certain future. To envisage this pixilated urban modernity, developers employed architectural images and computer simulations to enable planners and later potential customers to walk through the proposed developments. Hence, we could consider the enclave as a speculative, globally circulating model based on the "aesthetic typology of urban fantasies" (Cardoso 2016, 96). But increasingly, the fantasy is being rendered concrete as developers plan and construct and these spaces. At least twenty-six proposed new cities[1] in nine countries are scheduled to be built, incorporating more than 270 km^2 of enclaved space. This remains a selective geography, with the presence of enclaves reflecting concentrations of wealth in extractive economies such as Nigeria or the Democratic Republic of Congo (DRC), or those with middle-income status such as Ghana. Hundreds of smaller gated community developments that have been under construction in cities such as Accra for several decades will also join them (Grant 2009; Morange et al. 2012).

I estimate the development of these new city projects to require tens of billions of dollars of financing, with the *Africa Report* suggesting more than USD 100 billion in end values (Lutter 2019). For instance, developers of Centenary City, Abuja, and Tatu City, Nairobi, both claim potential final values at more than USD 18 billion; Konza Technology City, Nairobi, USD 16.5 billion; and Eko Atlantic, USD 6 billion. The developer, Rendeavour, claims up to USD 250 million of investment in the land, infrastructure, and construction of each of their six sites. Most of these enclaves require the assembly of complex financing agreements involving multiple partners. As an example, developers suggested Nkwashi in Lusaka will have a final value of USD 1.5 billion, and Thebe Investment Management, the project developer, has secured financing from Stanbic, Citibank, Madison Life Insurance, and Barclays Bank to fund the underlying infrastructure and multiple stages of development.

These vast sums of finance show that after decades of being on the margins of the global real-estate market, financing for urban projects is now flowing into large, economically growing cities at a significant scale. The

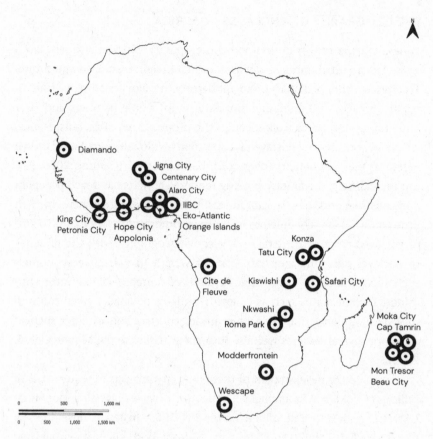

3.1 Map showing new cities proposed across Africa.

construction of Tatu, Konza, and others will dramatically shift the underlying urban experience for many of the upper middle class, and elite, and those who cannot access these spaces. I estimate a total population of more than 2.5 million residents and the promise of employment for hundreds of thousands in these developments. To illustrate, developers of Diamniadio outside of Dakar envision 50,000 jobs and 300,000 residents, and the developers of Konza Technology City estimate 100,000 jobs and 150,000 residents.

If these figures remain a small proportion of the urban populations of the region, and since the urban majority find housing and infrastructure beyond these developments, why do enclave techno-environments matter? Primarily, we should recognize these geographies as reinforcing urban inequality through uneven access to infrastructure. It is not a neutral process, but rather reshapes the social composition of the city to insulate the elite from the impact of infrastructural insecurity. This is concerning because this eco-segregation is gathering pace in an era of climate crisis and only seems likely to intensify.

ENCLAVES IN ACCRA

INFRASTRUCTURAL CATCH-UP IN THE WIDER CITY

Accra has been at the forefront of capitalist urbanization in West Africa for more than two decades in a country that is now classified by the World Bank as middle income and therefore an increasingly attractive site for real-estate investment. Richard Grant (2009, 7) explained that "The combining of global and local processes is producing a globalizing city: a production that is evolving and far from complete." It is this dynamic, difficult-to-decipher spatiality of the metropolitan region of four million that Grant terms "truncated modernization: the evolving city is fragmented, chaotic and spatially messy" (7). This generates a focus on thinking about what type of "modern city" is being assembled through the planning and con-struction of these urban enclaves on top of the messy, ever-shifting geog-raphies of Accra.

The promised infrastructural security of the enclave is in stark contrast with the incremental, makeshift systems of informal settlements or the mundane housing of sprawling suburbs. Here, a focus on the expansive edge-city spaces of places such as Kasoa beyond the western boundaries of the Greater Accra Metropolitan Area (GAMA) best conveys why the enclave has become popular. In order to get a sense of infrastructural life in Accra beyond the enclave and one of the key drivers in the growth of these spaces, it is necessary to understand the shifting geography of the metropolitan region. A senior municipal official in Kasoa told me that households "are moving towards the outskirts of Greater Accra. We have to deal with the pressure from the city." This growth beyond the official boundaries of the GAMA incorporates the adjacent Awutu Senya East Municipality, which has grown significantly over the last decade because of its proximity to

3.2   The sprawling landscape of Accra.

3.3   The infrastructures of Kasoa.

Accra. Like other towns and small cities, Kasoa, the primary town in the municipality, experienced rapid urbanization and is known as one of the fastest growing towns in West Africa. Its population surged from 35,000 in 2000 to more than 135,000 by 2021 (Ghana Statistical Service 2022).

This extended urban landscape of Accra's western periphery is dominated by the visibility of the newly installed infrastructure. Telecommunications towers are ever present as various operators race to ensure they will have the carrying capacity for the surging data needs of new population centers. Power distribution hardware at various scales is present, if not always completed or operational. At the municipal offices, larger pylons that carry electricity to western Ghana stand domineeringly over office buildings as localized networks of electricity, which are often interrupted, travel underneath. Water towers—large and small—dot the area, suggesting that the main water network is lacking in reach or capacity. They are situated next to people's homes and larger institutions, such as schools. In this infrastructural landscape, there is little time to react to or plan for the systems that are required by surging new populations. This dynamic results in a visibility of the infrastructure, layers of recent techno-history, some obsolete, some that authorities have built to keep up with populations pressures, which I term the infrastructural catch-up.

A combination of insufficient financing for networked investments in state utilities and intensifying (sub)urban growth (Mabin et al. 2013) produce the catch-up. Over the last two decades and mirrored across the region, Accra's expansion has reshaped peri-urban land into an expansive, dispersed urban form (Yeboah 2000), adding further costs to any infrastructural development plans. For instance, Ghana Water Company Limited, the utility company, admitted that it could not fully supply Kasoa. This forced urban dwellers to become dependent on off-grid means of securing water, such as through tankers that ply their trade in many neighborhoods (Alba et al. 2019). Distribution losses of up to 50 percent in the daily 700,000 gallons of water supplied to the municipality through existing pipelines compound the fractured provision of services. This is an issue caused by a lack of maintenance and repair (Town and Country Planning Department 2011). Although 30 percent of the population in Awutu Senya East can access piped water, about 50 percent of the population depend on informal provisions, especially water tanker services (which often cost twice the price of utility services) or local boreholes (Ghana Statistical Service 2021. The high demand for these informal services is because of the increasing population, ever-growing tariffs, and the lack of a centralized water network.

With the intensifying climate crisis, these under provisioned spaces have become vulnerable to untamed natures in which the existing infrastructure cannot cope. For instance, during widespread flooding in 2015, there was disruption to operations of electricity and water that lasted several frustrating days, while transportation infrastructure such bridges and roads had washed away or collapsed, in several instances resulting in fatalities. I was stood on a hill overlooking Malham Junction with Joe, a neighbor working as, among other things, a taxi driver. He told me he could not access the main road, which he needed to do to pick up customers. He lamented the capacity of various forms of infrastructure to withstand heavy rain and the myriad of problems his household would face in the proceeding days.

It is this patchwork of infrastructural provision, repeatedly failed attempts at infrastructural catch-up, the experience of disrupted services, and the growing specter of climate crisis that has driven customers to buy their way into the techno promises of the enclave.

NEW CITY ENCLAVE: APPOLONIA

Appolonia, "City of Light," is one of the largest master-planned projects in Accra, and it illustrates how enclave spaces run as quasi-independent private municipalities are being built in urban Africa. It is within the GAMA, 20 km northeast of the city itself and next to the existing towns of Afienya and Oyibi.

A global equity company, the Renaissance Group, delivered by the African development business, Rendeavour, financed Appolonia. The Renaissance Group was established in Russia in 1995, working with Credit Suisse on privatization, and has diversified as a company. The company generated its wealth by trading various assets, including investments in the African-wide Eco Bank and some 600,000 acres of forestry in far eastern Russia. Here, surplus capital generated by the company through the extraction of global resources attempts to find new accumulation opportunities through the financing of various speculative urban development projects. On the company website Rendeavour (n.d.) claim to be working on "30,000 acres of visionary projects in the growth trajectories of large cities in Kenya, Ghana, Nigeria, Zambia and Democratic Republic of Congo" and claims to be "Africa's largest urban land developer," with the master plan developed, I was told by "Tim," by a team that "are all ex senior city officials from Johannesburg."

Rendeavour claims investment of hundreds of millions of dollars to work on the land division, infrastructure, and management of Appolonia. The company purchased communal land from traditional authorities in Appolonia village—an acquisition that conveys shifts in the treatment of land in the region. Traditional land has lost its communal use, become commodified, and been sold off for urban development (Bartels, Bruns, and Alba 2018). As Lena Fält (2019, 444) described, while traditional authorities agreed to the sale, there were disputes and contestations from other actors within the area to the terms and conditions. If this wave of construction is benefiting some, there are many who are excluded and marginalized from the perceived benefits. The likelihood of tax reductions or exemptions is high for Appolonia, which would join other such projects in failing to contribute to public coffers.[2] According to Fält (2019, 454), negotiations were ongoing, with the developer insisting that reductions can be "justified since the developer funds all infrastructure within the new city."

Appolonia illustrates the convergence of global capital, rapid urban growth, and changes in the treatment of land. Concerns for sustainability and infrastructural security that are emanating from politicians, NGOs, and other actors set the conditions for its underlying vision. Rendeavour announced that "We aim to help create the infrastructure . . . that will help sustain and accelerate Africa's economic growth, meet the aspirations of Africa's burgeoning middle classes, and serve as a catalyst for further urban development" (Appolonia, n.d.).

Encompassing 800 hectares, the site itself should house 100,000 residents. Developers describe Appolonia as promoting "world class environmental integrity and sustainability" as part of a premium (eco) space (Appolonia, n.d.). "Tim," a member of the development team, told me, "We have a strong focus on being as sustainable as possible. Given the impact of what we do, I mean, if we develop land, you can have a positive or negative impact." The enclave will integrate so-called eco-innovation by circulating

3.4 Entering the enclave of Appolonia.
Photo credit: Lazarus Jambadu

international knowledge and innovations into sustainability, design, and technological operation. Appolonia allows the elite to bypass the infrastructure problems and unstable environments of Accra and promises a new model of sustainability. Rendeavour has promoted this new city to potential buyers as a means of infrastructural security that is not guaranteed in the sprawling suburbs and informal settlements beyond its walls—a techno-environmental fix for those able to afford to pay for its premium services.

Commenting in the *African Review of Business and Technology* (2012), Stephen Jennings, the Chief Executive of Renaissance Group argued that

"Ghana's powerful combination of a dynamic economy and business-friendly environment will allow it to create solutions to rapid urbanization and demographic growth faster than many countries in the world." This eco-boosterist rhetoric outlined a future in which enclave developments deliver sustainability and "world-class" infrastructure. Precisely what happens to the rest of Accra is unclear. Does Appolonia prefigure the techno-environmental remaking of the entire city? Or is it simply that the infrastructure solutions offered by Appolonia are part of the annexation of urban space to assemble archipelagos of wealth and premium technology, separated from the infrastructure of the outside through walls, gates, and security guards?

Promises of infrastructural security unavailable in the wider city underwrite the privatized offer of Appolonia. Energy supply is a particular issue in a city of constant interruption, and the development promises a "modern" experience of always-on air conditioning, home entertainment, and broadband. The backup supply integrates 18 km of new power lines connecting to the mobile bulk supply point and is supported by a 60 MVA–33/11 kV substation that was completed on site. Further off-grid energy infrastructure will be installed by Axcon Energy, leading to the operation of a 5 MW solar farm to be expanded in the future. Tarred roads with pedestrian walkways and covered, sustainable u-drains will feature throughout. Fourteen kilometers of dedicated water pipeline from the new Ghana Water Company Ltd. booster station in Kpone will direct potable water into the site. In addition, there are plans to cut leakage through deploying dedicated high-density polyethylene pipes for water and sanitation.

New, smart safety measures and high-tech security also feature prominently, alongside state-of-the-art ICT networks and the provision of ultra-fast internet services through fiber infrastructure. This "smart" security and connection system incorporates a data center that the operators, MainOne (n.d.), claim is "world-class infrastructure" of "private data center suites, enterprise-grade 24×7 multi-level security and video surveillance, precision cooling, safety and fire suppression systems." In homes, the developers offer the potential for a controlled environment by promising complete order over their climate through air conditioning and water heater preparation. Using Hydraform bricks in some buildings may offer a passive cooling alternative to air conditioning. Appolonia Connect, a private entity, manages and operates the city's utilities of power, water, and telecommunication infrastructure. This privatized utility management is a key part of the offer of what CEO Bright Owusu-Amofah suggested would be a "provide a world-class utility service" (Rendeavour, n.d.). Collectively, these private infrastructures of the enclave should facilitate control of technology and securitization of the environment.

Appolonia may help reduce citywide demands on the overstretched energy system through the creation of post-networked spaces and secondary infrastructures beyond the grid. It may also provide infrastructural security for its residents. Because Appolonia is managed under privatized governance, there is a price for living in the enclave that restricts those who can and cannot gain access to it. As the developer explained to me, "A large part of how we go about the master planning process is to set up quasi municipal authorities which you pay a fee for the built environment maintenance. From day one we make sure it is self-financed . . . to ensure the principle that you pay for the service you get." If the privatized space of Appolonia may appear to suggest the state's absence in the enclave's creation, the Ghanaian Government has played an important role in restructuring urban laws, policies, and everyday governance. This is an example of an entrepreneurial state (Mazzucato 2011) facilitating the process of private-sector investment.

In order to manage the new population in Appolonia, the developers have fully integrated infrastructural life into the latest technological advancements. Rendeavour's quasi-independent municipal authority functions with pay-as-you-go services and biometric technologies that they have built into homes. This includes the potential of locking out owners if they do not make service payments on time. As "Tim" alluded to, "If you can't pay your electricity bill, then you can't get in your place. This is getting people to be responsible for what they consume." Thus, new smart technologies function as disciplinary tools that mediate access to premium network services for residents. Buying a home in Appolonia is not enough to guarantee infrastructural security because the enclave is a privatized context in which the developer-manager securitizes their investment by ensuring that financial compliance is hardwired into the infrastructure. Much like residents in poorer districts with pay-as-you-go meters, households may find themselves cut off from essential everyday resources if they default on payments. From privacy concerns to tax exemptions to the lack of public oversight and the rollout of surveillance technologies, Appolonia offers an example of the problems that the privatized governance and quasi-independent municipalities of African new cities are generating.

Homes start at USD 86,599, with Rendeavour claiming this represents affordable housing in "an inclusive new city." The developer also offers plots of land for buyers who would prefer to build their own homes, with prices starting at USD 17,500. This remains far beyond the financial threshold for most residents in Accra, a city wherein the mean annual income per capita was USD 950 and the median per annum was USD 730 in 2012 (Fink, Weeks, and Hill 2012). The developer expects customers buying plots of land to build the homes quickly. This is in contrast to the piecemeal, years-long process that characterizes Accra's sprawling suburbs and allows households to finance construction. Appolonia's so-called affordable prices show how the developers have designed the enclave as a space for the few rather than the many. "Tim" argued that whether "you're poor or middle class doesn't mean that you should live in a bad environment, but it comes with a fee and it's part of our design process that this fee is embedded in all that we do." However, the cost associated with living in Appolonia makes it unlikely that the urban poor will enjoy these types of premium networked services. Indeed, this techno-environment is liable to fracture the already fragmented infrastructure of the wider city further.

MUNDANE ENCLAVE: FORTUNE CITY

Appolonia is the most spectacular of the master-planned enclaves in Accra. But it is not the only type of enclave techno-environment being configured. Mundane developments, such as Fortune City, are also reshaping techno-environments. This suburban enclave is situated down from what was still a dirt road in 2016 and behind the West-Hills Mall in the western suburbs of the GAMA toward Kasoa. Despite making no substantial contribution to the public purse to improve local infrastructure, Fortune City relies on various types of public investment and technology for connections to the city and its resource flows. As I traveled down the bumpy road to visit the showroom, the developers were still trying to convince the municipality to take on the costs of building and maintaining the

3.5 A model of Fortune City in the showroom.

road to their site, showing how enclaves rely on public subsidy for private profit. Out of the overgrown, peri-urban land, a high-walled, gated suburb appeared, signaling its separation from its immediate surroundings. The ubiquitous swimming pool and tennis court soon came into view as the imposing gates opened.

At the sales office, a replica model welcomed potential buyers to the vision of a self-contained community. The developer told me that Fortune City was to be targeted at the Ghanaian elite rather than expats or international investors. The ninety-seven houses planned were all under construction during its initial stage, and at the time of my visit, the developers had welcomed the first twenty-two buyers on to site. Fortune City is an example of the new residential enclaves that are scattered throughout the western periphery of the city. The development is 100 percent owned by Sunda (GH) Real Estates, a Chinese company with headquarters in Guangzhou. Sunda Group has set up various subsidiaries across Africa, with an initial focus of building materials that included tiling and PVC panels and everyday household products. They established the real-estate arm of the group in Ghana in 2011 and signified the first attempt to capitalize on the surging land prices of the region as it constructed and then sold four-bedroom homes with prices upward of USD 435,000.

As I walked down the newly laid and pristine roads, I noticed that the infrastructure differed from the surrounding homes that lay beyond the high walls. Streetlights were solar powered so as not to rely on interrupted supply. The developers were keen to inform me of the layers of infrastructural security they had established. The streetlights were only the most visible of a series of systems designed to provide off-grid backup and included a gas-powered system to generate electricity. The developers conceived of the off-grid solution as a longer-term (and quieter) alternative to the ubiquitous diesel generators that hum into action across the city during power interruptions. There were also plans for a reservoir to ensure a supply of water for homes, and as I was told, the communal swimming pool would be available at all times. These backup systems featured in the sales brochure as important features in the overall offer. During my visit, the developers wanted to impress potential buyers with this infrastructural security and its sanitized and manicured environment. The sales pitch clarified that "Fortune City seeks to create an elegant and unruffled landscape that assures our customers close interaction with nature for their maximum relaxation. The beauty of our tropical trees and flowers amidst the lavish style, gardens, fountains and artificial water bodies makes the city homey."

Fortune City is a very different enclave from Appolonia. However, its internationally financed construction and privatized governance, alongside the promise of always-on power, water, and other services as well as controlled environments and a suite of new technologies, show that the importance of infrastructural security in the making of these spaces, despite their differing forms.

LUXURY ENCLAVE: PEARL IN THE CITY

If Fortune City is typical of enclaves remaking peripheral Accra, we can find a third type of enclave techno-environment in the volumetric transformation of central parts of the city through high-density apartment schemes. In 2016, I visited a then under-construction development in Cantonments, an older residential neighborhood that housed elite and

3.6 Construction continues at Pearl in the City.

middle-class households near the massive, fortified US embassy. After being invited for a Turkish brunch by the friendly developer Erdinc, I took a short tour of the complex to gain a better comprehension of its technological configurations as it neared completion. Erdinc told us about their efforts to build 150 apartments of what was previously one colonial-era house that a British governor had reportedly used in days past. Bureaucratic difficulties of the Ghanaian land system had delayed the project (see Kasanga and Kotey 2001). But the first apartments were nearly ready for occupation, with prices starting at USD 120,000 for a studio and going up to USD 600,000 for a four-bedroom unit. The apartments, which had sold out by 2019, were of high specification and very different from those that had been built in Accra in previous decades. They appealed, according to the developer, to customers with aspirations for a global, modern, and luxurious lifestyle.

We moved to the car park, where a large transformer was situated. The transformer had a switch that could move effortlessly from the public electricity network to four backup generators, guaranteeing infrastructural security for residents. Solar panels were to be installed on the roof in order to generate heated water. Double-glazed windows were being installed to enable a specific microclimate in each apartment through control units that were placed in each room. Erdinc viewed these features as integral to the success of the project and in "doing things the right way," which he mentioned again when we passed the old trees he had retained. He argued that the benefits of sustainability were being recognized because of the financial savings for developers and owners and the capacity of these spaces to address infrastructural insecurity. There is an interesting subtext here: the environment forms an important part of the promotional material, like with Fortune City, promising the "privilege of a healthy and natural life" and the opportunity to "breath the clean air" that makes "Pearl in the City a special world custom tailored to you . . . at the epicenter of Accra, avoiding excess traffic and supported by excellent infrastructure, one may feel like they are submerged in a tropical Island."

The preview unit was nearly complete when we entered, and the space was a hive of activity. Electricians were installing energy-saving LED lights into the walls and control boxes that would allow residents to regulate microclimates in their smart homes. Outdoor shading was being installed to provide shelter from the intense sun—a somewhat rare feature in housing in Ghana. Most of the buyers included US- and UK-based Ghanaians who were looking to, as the developer explained, "invest back home" or for a convenient place to stay during visits before a potentially longer return in the future. Before completion, the developer had already sold 40 percent of the units, which was still a relatively rare occurrence in Accra.

The enclaved techno-environments of Accra, whether Appolonia, Fortune City, or Pearl in the City, envisage a particular urban future. Infrastructural life is delineated by those who live on the inside and those who live on the outside. These cases draw attention not just to how the city is being remade but to how this remaking may have wide-ranging implications for the provision and planning of public infrastructure. This is especially important in considering revenue streams for municipality as elite/ middle-class customers opt out of various payments that are diverted into the companies running these privatized spaces. If the enclave represents a particular form of modernity, we have to ask what is left behind for those who depend on the city and its public infrastructure outside these privatized archipelagos?

ENCLAVE POLITICS AND IMPERATIVES

What do the techno-environments of enclaves in Accra tell us about Africa's urbanization, and how do these dynamics relate to the Infrastructural South? Enclaves are a negation of the representations of urban Africa as a space for substandard infrastructures, deficits, and supposed technological backwardness. Appolonia, Pearl in the City, and Fortune City all show Accra at the cutting edge of experimentation in new forms of sustainable urban living and on-demand, smart services that convey the mutating techno-environments of urban Africa. This infrastructurally secure enclave model is not derived from the West. Rather, it challenges the teleological narrative of the urban experience and makes visible how the universal notion of modernity transforms through relational processes, the transfer and circulation of knowledge, finance, technologies, and architectural forms, and in the encountering of the particular time and space of cities such as Accra. The model of the enclave undoubtedly owes much of its makeup to the mobilization of policies, models, and architectural forms from Asia and what Watson (2014, 222) terms the "iconic eastern cities." However, these techno-environments have also materialized from the variegated conditions across the region. Indeed, these geographies illustrate the unique infrastructural pathway upon which urban Africa is traveling—one that is at once shared and distinct from experiences of infrastructural modernity in both Norths and Souths.

Because enclaves in urban Africa bear little similarity with techno-environments that have come before, there is an imperative to consider the urban future being configured. Even as developers transform gated communities with longer histories because of concerns with infrastructural security, the enclaves of new cities, suburban spaces, and the remaking of inner cities remain in their infancy. The enclaves explored in this chapter show these transformations are intensifying. This proposition leads us to two potential futures: a future in which enclaves pioneer sustainable technology and experimentation that tests, pilots, and configures new infrastructure operations and may be rolled out to the rest of the city, address resource demands, and improve quality of life, or a future where enclaves result in a form of eco-segregation. This is a division between those who can access these spaces and their premium services and those who cannot. Whether an informal settlement or a mundane suburb, communities may be left facing a series of infrastructural problems that were clear when I stood with Joe on the hill overlooking Malham Junction during the floods of 2015.

SUSTAINABILITY AS ELITE SECURITIZATION

Promises of infrastructural security are central to the making of enclave techno-environments. Such assurances are celebrated as being able to support broader efforts to address infrastructural challenges across the region's rapidly urbanizing spaces. Attempts to establish forms of infrastructural security may provide opportunities to pilot, experiment, and transfer new efficiencies, technologies, operations, and user experiences to public networks. The urban experimentation of the enclave may hasten important transformations in infrastructure planning and operations. To clarify, municipalities may take up experiments with smart technologies that can control household environments in Appolonia and roll them out to the public network. Other construction companies could start employing the Hydraform bricks besting tested. The LED lights and solar water heaters in Pearl in the City could become standardized in non-premium projects as developers realize their financial benefits. In addition, Appolonia's developer suggested that the off-grid networks being constructed could contribute to more efficient operations in public infrastructures. Although the potential for energy-generating or rain-harvesting technologies to feed back into public networks remains in a nascent stage of experimentation, such pilots keep the capacity to be taken up in the broader infrastructural landscape (Castán Broto and Bulkeley 2013). However, this optimistic position misses the contradictions that lie beneath enclaved promises of sustainability. Masses of concrete underpin most of these urban developments. These enclaves join the pervasive concrete shells of the self-built homes that have transformed the peripheries of African cities. Builders used concrete blocks in 78 percent of housing across Accra (Acquaah-Harrison 2004), which is perhaps the foremost material of the third wave of urbanization. Builders use concrete because it is relatively cheap and readily available and because it allows for architectural details that symbolize an (imported) middle-class/elite aesthetic (e.g., classical columns, arches, etc.). Cement consumption is increasing faster in sub-Saharan Africa than anywhere else, outpacing even China, with growth in production at 6.5 percent in 2014 and an average of 7.2 percent between 2010 and 2017 (Ajimotokan 2018). This demand has predicated investments in increasing the capacity of cement production in various African countries. For instance, the Dangote Group, a Nigerian conglomerate, has invested more than USD 5 billion in new production facilities (Bah, Faye, and Geh 2018) to support the construction boom in ten countries that include Ghana and reported a USD 1 billion in profit in 2018 (Oji 2019).

The energy-intensive nature of concrete is significant because of its high volume of production and because concrete buildings require air-conditioning systems due to their thermal inefficiencies. This has led to increasing demands on the fragile electricity network. An Accra-based architect warned about the air-conditioning required: "In the concrete homes of the middle and upper classes, there is a reliance on mechanical means of cooling, and there is little space for these traditional techniques or motivation to do so." As I will reflect upon in chapter 10, these construction techniques turn their back on centuries of climate-influenced traditional design across the Sahel region, including the iconic buildings of the Sudanese style of architecture, such as Larabanga Mosque in Tamale. A developer in Accra told me, "Mud was seen as backward, and cement was seen as progressive. So, if you go to Keta for instance, somebody will say they have never lived in a mud house because their grandfather was rich. So, we lost the sense of the utilitarian value of how we use to build."

In creating these enclave techno-environments, failing to build with anything other than concrete shows that mistakes of recent decades have been or are being repeated in the next phases of city building in Accra and elsewhere. If urban Africa is undergoing a particular transformation, the materiality of urban modernity, concrete, remains shared with cities across the planet. It is worth questioning how enclaves interact with wider attempts to develop sustainable futures across the city because they are in effect drawing on the resources of the wider city, whether water or labor, land or electricity. The enclave enables securitization and sustainability but only for the few. The energy-intensive nature of constructing and operating enclaves means imposing costs to existing infrastructure and people that are shared with the wider city, while premium networks remain mediated through capacity to pay. This archipelago techno-geography predicated on the allure of infrastructural security comes at a cost for broader efforts at sustainability. The enclave also prompts us to think about its segregationist premise.

ECO-SEGREGATION?

The techno-environments of enclaves are establishing an eco-segregation— a marketized infrastructural solution to widespread fears and insecurities related to techno-environmental conditions. The surge of proposed developments with high-tech sustainable services is contributing to a post-networked infrastructure of the city. This is a geography where premium space becomes unbundled (Graham and Marvin 2001) and privatized from the partial public infrastructural network. Olivier Coutard and Jonathan Rutherford (2011, 107) describe these configurations as "'small scale', 'decentralized', 'dispersed', or otherwise 'alternative' technologies." What makes enclave techno-environments distinct in urban Africa from many other contexts is that it is being pursued in a context in which governments have not fully delivered basic services. For instance, only 45 percent of the population in Accra has access to piped water, and the supply of electricity for those on the network is consistently interrupted.

The enclave poses important questions about the infrastructural futures of the wider city and politics contained within these visions. It is important to reflect upon whether the evacuation of the incipient middle class and elite from existing city space to these enclosed, privatized spaces, and their subsequent abandonment of the everyday problems of the old city, will leave any political imperative or fiscal power to address various infrastructural challenges. The promise of controlled environments, fully functioning infrastructural services, and a perceived global, modern lifestyle enable sectors of society to opt out of broader urban political struggles that would otherwise push the state to invest in public infrastructure. That this might signify a rejection of the politics of universal service and a shared responsibility for the future of the city is deeply concerning.

If, as promised by Appolonia's developers, this type of development is "Africa's urban future," then it is likely that segregation will be entrenched in Accra and elsewhere as it becomes integrated into the urban environments of the region. As De Boeck (2011, 277) argued about Cité du Fleuve, this spatial segregation echoes the colonial divisions of the city. In Ghana, this division was through the British government's reserved areas (Killingray 1986), showing the "imperial durablities" (Stoler 2016) and mutating forms of segregated urban space across time. A segregationist premise is particularly concerning because of the intensifying effects of climate change and its impacts across urban Africa (Lwasa 2010), pushing for a recognition of this dynamic as eco-segregation. With the onset of the climate crisis, the enclave will offer a protective built environment and may become one of the dominant spatial forms of the third wave of urbanization. It has become a material form of segregation pursued through a differentiated, class-based experience of nature and the environment, infrastructural services, and technology.

Eco-segregation resonates with Frantz Fanon's (1967, 39) writing on the rift between colonial-era "settler town" and "native town," which I outlined in the previous chapter, emphasizing Fanon's focus on "reciprocal exclusivity." As the climate crisis becomes ever more pronounced, the infrastructural topography of the enclave appears likely to offer a twenty-first century version of the spatial division that has mutated from its colonial form vis-à-vis circulating models of new cities from the South. By fencing off the rich from the poor, access to infrastructure becomes more splintered and fragmented. However, the enclave has its limitations for those seeking refuge from environmental turbulence. As Watson (2014, 229) argues, and extreme weather events or the coronavirus pandemic have shown, the capacity to completely "insulate themselves from the 'disorder' and 'chaos' of the existing cities is remote." These spaces demand we update Fanon's (1967) description of the settler town and native town. It forces us to comprehend better how life inside and outside of the enclave will probably reinforce and amplify techno-environmental inequalities.

Eco-segregation as integral to the making of such urban developments, alongside the push for sustainability and "world-class" infrastructure, may cause the displacement of existing communities in the name of climate resilience and infrastructural security. As an example, the mayor of Accra between 2009 and 2017, Alfred Oko Vanderpuije, mobilized public concern over climate change in response to the 2015 floods, wherein at least 225 people died, 8,000 lost their homes, and more than USD 100 million of damage to the environment was caused. This acknowledgment of climate change was used to justify evictions in Old Fadama, one of the city's popular neighborhoods and its largest informal settlement, with more than 35,000 residents. The state has long tried to evict households from this contested area. As Tom Gillespie (2016, 72) argues, demolition was planned "so that private-sector developers could build high-rise mixed-use development." Gillespie then discusses the implications of these evictions by noting that "the displaced squatters will be replaced by those who can afford to access the hotels, offices and apartments that will replace the shacks that currently occupy this land." A potential enclaved future for Old Fadama is envisioned to attract new circulations of international capital built upon architectural fantasy and discourse about climate risk (Afenah 2012). In the place of a home for thousands of the urban poor, there is a still yet to be a realized vision to establish a secure, segregated enclave of sustainability and high-income living where current residents have little chance of accessing it.

Disturbing proposals for so-called charter cities from neocolonial fantasists illustrate the reactionary futures that enclave techno-environments may enact through eco-segregation. The key logic behind charter cities is that they are built with the priority to manage the space, its people, and infrastructures while controlling governance and finances delegated from municipalities and governments. As I have explored, these arrangements fit neatly into the large, master-planned projects that are being proposed and support the potential for infrastructurally driven eco-segregation in the future. Enclaves may challenge national sovereignty, municipal oversight, and territorial integrity because the spaces they create are privatized and run by corporations that can bypass local tax obligations, laws, or standards, resulting in further division between inside and outside.

Advocates mobilize the techno-environments of enclaves as a shortcut to achieving a perceived urban modernity to make modern the African city through the creation of spaces with managed techno-environmental conditions. It is a model that has transferred its pretensions of modernity and being global from other urbanizing regions in the South, showing how modernity in urban Africa is both imported, situated, and configured beyond the typologies of the Western city. The costs of this shortcut will probably be high, for the enclave represents the withdrawal of responsibility to address the challenges of the wider city. It is a space of restricted access erasing the poor from its boundaries—despite the daily movements of cleaners, maintenance staff, and others into and out of the space. This segregated space that excludes will perhaps also be benefiting from tax breaks and fiscal incentives from the public. This techno-environmental fix establishes an infrastructural security for the few but abandons the notion of collectivity that cities have sometimes historically engendered. I have shown in this chapter how this might be understood as a techno-fantasy of start-again city making, wherein the task of delivering, operating, and maintaining infrastructure does not extend beyond those who can pay for their piece of real estate, and becomes a site of accumulation for real-estate finance. If new experiments in sustainable living, technology, and infrastructure operation may be established, these are accessible only to the few, and they are conducted at the expense of addressing wider infrastructural futures. In this era of rapid climate change, urbanization, and ongoing technological dysfunction, this chapter has demonstrated how the enclave delineates an experience of urban modernity between those who are able to buy their way in and those who are left to navigate the outside.

4

BETWEEN SURVIVAL AND THE PREFIGURATIVE

URBANIZATION FROM BELOW

The operations and geographies of infrastructure across African cities are a dense patchwork of hybrid technologies, systems, and relations sustaining everyday social reproduction through the circulation of essential resource flows. These are incremental techno-environments that center heterogenous, ever-shifting arrays of people and infrastructure. I suggest understanding the remaking of these networks as critical in the broader composition of urban Africa. This means that when focusing on the myriad of ways in which urbanization proceeds from below and how the urban technical is adjusted, remade, and reconfigured, a distinct urban modernity comes into view. This is a lived experience for the many that is so well conveyed in fiction such as Chris Abani's (2005) *Graceland*, which is set in the informal parts of Lagos. He writes, "Elvis opened his mouth to answer but thought better of it. The road outside the tenement was waterlogged and the dirt had been whipped into a muddy brown froth that looked like chocolate frosting. Someone had laid out short planks to carve a path through the sludge." In this short, sharp passage, Abani offers a concise snapshot of the everyday challenges facing residents and the tactics they employ to assemble infrastructure to navigate such terrain.

Incremental techno-environments defy neat definitions of formal and informal urban conditions. Emphasis on disrupting the where and when of life on/off the urban grid enables a clearer articulation of these overlapping, multiple systems, the urban politics wrapped up in everyday experiences of the network, and the ways a constellation of techno-environmental and social relations ensure urban dwellers can access resource flows. Sometimes, this is a great burden and at a cost to those involved. At other times, people find ways to make a livelihood or a side hustle, or they come together to prefigure a communal infrastructural operation. Incremental techno-environments remain highly symbolic, as well as being a material process through which representations of urban Africa frame it as unmodern—a representation of a vast zone of slums in infrastructural "deficit" that continues to require technological intervention and development assistance from afar. In this narrative, these towns and cities are supposed to mirror the historical experience of the Western city and its networked model as a pathway to attain a perceived universal urban modernity through a copy/catch-up approach. However, I suggest these infrastructural geographies are a distinctly modern phenomenon, as much as the networked spaces of the Western city mediated through the "symbolic importance of the modern ideals of integration and cohesion" (Graham and Marvin 2001, 42).

Life amid these incremental techno-environments is a mode of inhabiting the city through the unequal experience of urbanization, so well articulated in popular music such as that of the Cape Town hip-hop collective Soundz from the South, Senegalese duo JT Rappé, or Nairobi-based Kamwangi Njue. As the next chapter will pick up, this is the perpetuation of the underdevelopment and exploitation of sub-Saharan Africa (Rodney 1972) and the resulting urban conditions that shape the infrastructural present. It means that basic services taken for granted in much of the North do not appear in the same technical form, that is, a centralized, unitary system operated by a single actor (whether state utility or private sector). A critical underpinning of this book has been the rejection of African cities as unmodern, when assessed against the technological and environmental criteria of the so-called modern, Western city. As the artwork of Romuald Hazoumé or writing of Sundaram (2009) allude to, these relations situate urban dwellers, their knowledge, collaborations, strategies, tactics, mishaps, and successes as thoroughly modern. Infrastructural modernity in these contexts is experienced through constant adjustment with the multiple, shifting infrastructures of everyday life, producing a patchwork urban geography of service provision. For tens of millions, life in the city entails having to intervene in, grapple with, and remake the technologies of social reproduction in the absence of universal, fully operating networked system. This is an urbanization from below as residents, communities, and a plethora of informal (and sometime formal) service operators attempt to connect, control, accumulate from and access, and reconfigure and adjust the resource flows that make the urban experience viable, however unstable.

INFRASTRUCTURAL DEFICIT

The *Financing Africa's Cities: The Imperative of Local Investment* study (Paulais, 2012) established the need for a minimum of USD 30 billion per year to cover the mounting infrastructure services required across urban Africa. UN-Habitat (2020) estimated that up to 69 million African urban dwellers have no access to safe water services, with the highest numbers in Central Africa (22 million), West Africa (21 million), and East Africa (19 million). And only 44 percent of urban residents in sub-Saharan Africa have access to basic sanitation services (i.e., improved sanitation facilities not shared with other households). Access to hygiene facilities for urban residents is low, with only 37 percent having basic handwashing facilities in their homes—a concern heightened during the COVID-19 pandemic (Onyishi et al. 2021) but long a problem in attempts at managing pathogenic diseases outbreaks such as cholera (Chigudu 2019).

Across these unequal experiences of water and sanitation service provision, urban dwellers find ways of making life viable through using multiple configurations of technology, resource flows, and social relations that may allow for some form of safe water supply or basic sanitation service. These are often beyond the centralized pipeline networks that offer only partial coverage to urban populations. As a study by GIZ (2018) found, "Only 56 percent of city-dwellers have access to piped water, down from 67 percent in 2003, and just 11 percent to a sewer connection," demonstrating the challenges that lay ahead for municipalities, urban dwellers, utilities, and national governments alike.

As the UN-Habitat (2014, 41) *State of African Cities* report made clear, "The growth of Africa's energy sector is a prerequisite for sustained expansion in all others." And electrification remains a central and pressing infrastructural need across various policies and plans; from the African Union's Vision 2063 through to national economic strategies, and subnational, regional, and municipal domains. Yet, in countries such as Malawi, Burundi, and the Central African Republic, less than 15 percent of the population can access modern electricity networks (IEA 2021). If the urban electrification rate is higher in urban areas than in rural hinterlands, many urban households get by without full access to electricity. By 2018 across sub-Saharan Africa, only 75 percent of urban dwellers had access to electricity (World Bank, n.d.). If this rate remains higher than in rural hinterlands (with only 45 percent of Africans in total having access), it still belies how formal, centralized, and universal basic services remain unextended to many parts of the city.

Data on urban electrification varies across this diverse continent, with 93 percent of urban dwellers in Equatorial Guinea able to access electricity. In Uganda, it is 55 percent, and in Ghana and South Africa, it is 90 and 88 percent, respectively. Collectively, these statistics highlight that sub-Saharan Africa has the lowest global rate of urban electrification at 68.8 percent (SAMSET 2016). What we consider an infrastructural deficit in towns and cities in relation to urban energy remains significant. The symbolism of failed attempts to establish a universal electricity supply sustains the notion of the not yet modern African city. These statistical overviews do not account for those unable to afford electricity or experiencing constant interruptions, even if these households are classified as connected to the grid because "Millions of low-income households simply do not have enough regular income to buy (enough of) the electricity they may now have access to, forcing many to make tragic choices between buying electricity, water, food or clothing" (McDonald 2009, 16).

In place of an electricity supply, a range of so-called nonmodern energy sources and flows power everyday life for tens of millions of urban dwellers (Castán Broto 2019). At a household level, electricity powers only 14 percent of residential demand, with charcoal (48 percent) and wood (23 percent) accounting for the most significant sources (SAMSET 2016). These energy data challenge whether a linear transition to so-called modern urban energy services—that is, being connected to a centralized electricity grid—is possible or even desired. Amid the intense speed and sheer numbers of new urban dwellers across Africa's urbanization process, and with an intensifying climate crisis, it is important to question the logics and certainties of these centralized models of infrastructure deployment and operation that have gone before.

In the so-called informal spaces of urban Africa, the deficit is particularly pronounced, as these parts of the city remain un(der)electrified, and accounts for a significant proportion of the more than 100 million urban dwellers unable to access electricity in the region. Many people may live in proximity to the electricity grid, but for economic reasons (connection costs of more than USD 250) to technical reasons (the massive backlog facing utility companies), they may not access this energy source. Further issues pertaining to these informal urban spaces and the problem of extending the urban electricity grid include issues about recognition of land (often a precursor to formal electricity connections), concentrations of poverty in these areas (in sustaining flows of electricity), high levels of density that make interventions difficult to plan, and the health and safety issues pertaining to unauthorized connections, including electrocution and "shack-fires." Youba Sokona, Yacob Mulugetta, and Haruna Gujba (2012, 5) set out the implications of this lack of urban electricity supply, arguing that "The low levels and lack of access to modern energy services for productive activities has also impacted negatively on development and entrenched poverty in the continent." The imperative to think anew after decades of development efforts orientated around the delivery of centrally run urban electricity services becomes even more pressing.

As the chapter on imposition techno-environments will set out, from colonial legacies to the chilling effects of structural adjustment policy, this infrastructural deficit results from the underdevelopment of Africa through the legacies of colonial planning and subsequent neocolonial, World Bank, and IMF control. It manifests through a lack of readily available finance and state capacity to deliver and operate basic urban services, despite determined efforts by national and municipal authorities. These financial gaps and the exclusion of African cities from circulations of global finance persist, despite recent surging investment flows into the economic infrastructures of roads or railways (Goodfellow 2020).

Investment remains a fraught, difficult process for municipalities in ongoing struggles to raise revenue streams to deploy and operate infrastructure. Unequal investment geographies require the local state to connect to financial flows that are focused on national governments as the preferred beneficiaries of donors (Obeng-Odoom 2013b). The turn toward the private sector, intensified through ongoing World Bank–led reforms, especially concerning procurement from the late 1990s, have partially or wholly privatized infrastructure operations. This has more often than not been as a public–private partnership (PPP)—a standard outcome of neoliberal World Bank/IMF prescriptions. Broader fiscal environments and unequal global economic histories remain critical to understanding the making of incremental techno-environments across energy networks and other sectors.

INFRASTRUCTURE AS INCREMENTAL

Throughout much of urban Africa, residents navigate the demands of social reproduction with little help from the state through the "fleshy, messy, and indeterminate stuff of everyday life" (Katz 2001, 711). In response, households intervene in ever-changing configurations of infrastructure to connect to resource flows, including across water and waste, electricity, and sanitation sectors. In cities such as Accra and Kampala (and in contrast to larger-scale investment by either market or state mechanisms), residents are engaged in a series of entanglements with the material and the social that collectively establish a crucial if sometime contradictory force in the making of the Infrastructural South. Primarily, I wish to emphasize these geographies as relations between people, technology, and the environment—a constant, unfolding dialectic of adjustment and readjustment across what Edgar Pieterse and AbdouMaliq Simone (2017) term "make + shift" urban geographies. Achille Mbembe and Sarah Nuttall (2004, 369) argued, "Like the [African] continent itself, [urbanization] is an amalgam of often disjointed circulatory processes . . . it has become, in spite of itself, a place of intermingling and improvisation."

In thinking about urbanization from below as mobile, fluid, and always in a state of becoming, Mbembe and Nuttall's work helps to place infrastructure not as a static set of technologies but rather as an ever-shifting configuration of human and nonhuman actors alike. These struggles around social reproduction challenge the assumptions about who is involved in the governing of infrastructure in the city in which "the association of infra-structure with state-based political authorities, long the main providers and regulators of public works, is largely taken for granted" (Chalfin 2014, 106). These forms of governing and managing infrastructure belie the lived expe-rience of tens of millions of households across the region involved in daily intersections with the materialities, operations, and flows and circulations of infrastructure. In recent years, analysis has reached beyond the role of the engineer or policy maker to think about how an extended range of actors remake these urban networks (Alba et al. 2019; Baptista 2019; Cirolia and Scheba 2019; Munro 2020; Schramm and Wright-Contreras 2017).

The incremental is a dispensation that addresses problems of access and cost, making possible flows of energy into households, or creating new fresh water supply arrangements. The motivation of these interventions is to channel resources toward the living of sometimes, though not always, marginalized urban lives as the basic platform of social reproduction in the city. Incremental urbanism exists at the periphery of city planners' official, mapped, and regulated networked geographies, even as it remains the foremost experience of urban modernity for the many. It is in these in-between spaces that the urban majority generate an alternative concep-tion of infrastructure, making adjustments to material and social relations both within the home and the wider neighborhood, on and off the urban grid, sometimes with success, and other times leading to failure, harm, or financial burden. Through incremental techno-environments, various individuals and collective social formations work hard not only to sustain survival in the city but also to generate opportunity and livelihood—from the self-employed toilet entrepreneur running their own self-build facil-ity to the companies bringing fresh water in tankers to the neighborhood to the community-run solar scheme providing much needed light in the evenings. This is a popular economy of infrastructure forged out of the gaps in service provision left by the state.

There are several important conceptualizations of the incremental,[1] which I characterize through two analytical imperatives. First, such work has sought to describe the existing infrastructural geographies of much of urban Africa, including the coexistence of centralized, formal systems, and the dispersed, less formal operations running in parallel, overlapping, sometimes working in harmony, and other times in opposition. A series of generative concepts have helped to explain these techno-environmental conditions. These terms include hybridity, suggestive of the particular ways in which urban dwellers access, experience, and pay for infrastructure services across formal and informal networks (Guma, Monstadt, and Schramm 2019; Furlong 2014; Jaglin 2014). And Sylvy Jaglin (2015) goes further, suggestive of African cities as post-networked spaces because these hybrid infrastructural geographies are functional through the myriad of ways in which they can deliver services beyond the centralized network and that show a viability and alternate way through which resource flows circulate through urban space.

Other articulations focus on the diversity of infrastructure through the term heterogeneity to highlight and make visible infrastructure beyond the centralized but partial networks operating in the Infrastructural South. Collectively, a growing number of studies focused on different sectors have emphasized the multiplicity of technologies, relations, economies, spaces, and temporalities in how urban systems operate (Munro 2020; Rusca and Schwartz 2018; Smiley 2020). As Jaglin (2014, 445) argued, "The model for integrated utility services has ignored the heterogeneous character of real delivery configurations, while maintaining the fiction of services based on national and egalitarian standards." In a co-authored paper with colleagues (Lawhon et al. 2018), we pushed for a heterogeneous infrastructure configuration framing. This way of thinking through infrastructure moves it from a thing or object toward a geographically dispersed set of configurations, all of which have a myriad of technical operations, social relations, types of risk, and power relations and are constantly shifting across everyday usage. This term goes beyond that of hybridity by thinking beyond the formal and informal because it "captures not simply the mixing of two (or more) kinds, but that the kinds being mixed are not clear from the outset" (Lawhon et al. 2018, 727).

Second, theorization has focused on the (re)making of infrastructure beyond or at the margins of the formal, centralized networks—the lived experience of the city for the urban majority. These studies have articulated several useful concepts that help to decipher the production of infrastructure not just through state investment or utility company plans and actions but also across the everyday worlds of the Infrastructural South.

The production of infrastructure occurs through everyday actions, sometimes leading to collapse, breakdown, or closure, at other times, requiring further investment, intervention, or collaboration. Incremental techno-environments are in a permanent state of flux. These conditions reflect the broader trajectories of African urbanization and the lives of households in rapidly urbanizing spaces. Nothing is stable, established, or closed. Studies recognize the incessantly shifting configurations in which people seek to intervene in operations and become entangled in the making of techno-environmental relations in which "infrastructures are always precarious achievements and service delivery is always a process in the making" (Baptista 2019, 510). Studies have highlighted the importance of examining the microscale of operation, the need and capacity of urban dwellers to interact with infrastructure, and the various social relation and politics that are produced.

A focus on state and utility actors is necessary but should not always place these intermediaries to the fore, nor consider the plans, operations, and investments to be more static than those of the urban dwellers engaged in the transformation of techno-environmental conditions. Indeed, these formally planned infrastructures, authorized and built by the state and private sector, operate in highly unequal ways and often sometime alongside less than formal systems of how (for example) policies and legislation become enacted. For instance, Cecilia Alda-Vidal, Michelle Kooy, and Maria Rusca (2018) examined how experiences of water supply in Lilongwe were, even on the centralized network, differentiated across the city because of the everyday decisions and calculations made by water engineers.

Formal actors are integral to the incremental. The need to consider the labor, collaborations, and actions of those intermediaries operating on the margins of the state and definitions of legality are also important. These are actors involved with the less official network spaces of the city, with no or only partial service provision provided by the central system. Of particular importance in bringing these actors to the fore of scholarship has been Simone (2004b), most prominently through his idea of "people *as* infrastructure." This concept extends "the notion of infrastructure directly to people's activities in the city" and, in doing so, "emphasizes economic collaboration among residents seemingly marginalized from and immiserated by urban life" (407). From informal providers distributing water supplies into unpiped neighborhoods to children carrying out household waste, this idea of people as infrastructure has become a powerful way to demonstrate the relations between people, things, and materials. It shows how people enroll themselves through either necessity or opportunity at the end of the network. Thinking through people as infrastructure generates a related term: social infrastructure. These are the myriad of associations and connections that people develop with each other that make up what Colin McFarlane and I (2017, 463) have described as the "flows, movements, congestions and internments of people and things" that enable urban life to be viable in resource-poor contexts.

Both the notion of people as infrastructure and the associated term of social infrastructures have highlighted the importance of situated understandings (Lawhon, Ernstson, and Silver 2014) of infrastructural intermediaries that begin from the everyday conditions, experiences, adjustments, inequalities, and potential held within these diverse social relations implicated in the operation infrastructures. As Sophie Schramm and Basil Ibrahim (2021) argued through a study of water in a middle-class Nairobi housing estate, the diverse experiences and operations of infrastructure, and the intermediaries involved, can vary from household to household, including far beyond the so-called informal and poor dwelling spaces of the city.

HYBRIDITY AND HETEROGENEITY

Namuwongo is a popular neighborhood in central Kampala, with more than 10,000 residents, next to the industrial zone and the wetlands and alongside the railway leading into the central city. With many facing ongoing threats of eviction and demolition, residents live in a range of dwellings and engage in a series of different energy practices, linking up to different energy generation and access sources and undertaking many usages. Many of the residents in Namuwongo are poor and struggle for everyday essentials, including fuel for cooking and light for the evening.

In research undertaken with Joel Ongwec on household energy use in Namuwongo, a range of different networks were operational across the neighborhood that make clear both the existing geographies of incremental urbanism and the lived experience amid this ever-shifting infrastructural topography. These included legalized and less than legal electricity networks,

4.1 Namuwongo, a popular neighborhood close to central Kampala.

firewood and charcoal primarily for cooking, and various makeshift technologies for lighting. Collectively, these networks help to power everyday
life in which households are located in a dense infrastructure topography
that plays out through the hybridity of different flows that need accessing,
connecting to, and using. Two key factors help to establish the explanation
for why hybridity and heterogeneity characterize the infrastructural geography of the neighborhood.

First, there is a lack of universal coverage of the centralized, formal electricity network across the neighborhood in parts of Namuwongo. Electrification rates in the surveyed households were just over 70 percent. However,
even in households with a connection, this was being supplied through
heterogenous means but dominated by either official connections from
Uganda's main electricity distribution company, UMEME (48 percent), or
from an informal connection from landlords (41 percent). The actually
existing conditions of the electricity system of Namuwongo destabilize the
binary construction of formality and informality because they show both
classifications at play and simultaneous, often in adjacent households or

even within the households themselves. They also highlight the myriad of intermediaries involved in supplying this energy source. Much of the neighborhood is now served by the utility operator. This does not go, at least officially, to the areas with the highest levels of informality, such as Soweto or the houses that lie close to the wetlands or railway land. The reason for this lack of network coverage is that UMEME, Uganda's largest electricity distributor, like utility companies across the region, is legally prevented from servicing these areas.

The boundaries of legality and authorization imposed by Kampala Capital City Authority (KCCA) and Rift Valley Railways (and now controlled by the Uganda Railway Corporation) through seeking to control the land they own in Namuwongo (including through eviction and demolition) become a seemingly impassable boundary to installing a new electricity infrastructure. But this does not stop the growth of the network beyond the reaches, guidelines, and mapped boundaries of the state. Instead, landlords, assisted by rogue electricians, use official connections in authorized/legal areas of the settlement to establish new circuits of electricity to these households. Wires hang in every direction from the larger and more permanent buildings to those that may be informal and precarious, in both construction and status. Charging a flat fee for usage of the power of up to 30,000–40,000 shillings (USD 8–10) per month, landlords become secondary suppliers. Households tend to limit their electricity usage, in these now-connected spaces, to lighting or charging of phones. Like other service provisions, this flow provides an important life-support system for the household to go about their everyday lives.

The materialities of formal and informal electricity networks sit alongside each other in a relationship of hybridity, overlapping and brought together through the needs of the households, the accumulation opportunities of the landlords, and the skills and capacities of local electricians. Thus, the meter that records the electricity of a landlord also covers the unauthorized connections that visibly lead out from an official connection to other households. Speaking to an official at UMEME, he estimated up to thirty households in Namuwongo could use one official connection. In these instances, the landlord has become an informal power distributor alongside the utility company, supplying customers, issuing bills (of sorts), taking payments, operating and maintaining the network, and adjusting prices in line with the rises instigated by the utility company or perhaps other calculations. Here, we must go beyond the technological hardware of the electricity network to see how the management of these infrastructures can also destabilize the binary between formal and informal, bringing to the fore the notion of hybridity and creating new power-imbued relations.

The second key factor for this infrastructural heterogeneity is ongoing energy poverty among many households in Namuwongo. Stefan Bouzarovski (2014, 276) helpfully defined energy poverty as a "situation in which a household lacks a socially and materially necessitated level of energy services in the home." In our survey of 100 households in early 2015, we used this definition of energy poverty and found 95 percent of households in such a situation. Households also expressed concern that energy was becoming more expensive. Energy poverty is a serious burden to household finances. Residents in Namuwongo struggle with the costs of energy in daily life and the calibrations of household budgets when resources are tight. This causes the need to have different options for powering everyday life—a mode of survival in the contemporary energyscape beyond simply plug in and power.

Urban dwellers in Namuwongo, as elsewhere, have to make economic trade-offs to secure energy supply. "Josi," a young mother, informed us, "We cannot afford medical bills because we use all the money on energy," and she suggested she would have to "go hungry." "Steve", at the Ministry of Lands, Housing, and Urban Development, highlighted this challenge of balancing and adjusting various household budgets. He suggested, "The survival mechanism of the urban poor is that they rely on borrowing credit, give me a little bit of this and pay later, now it is first pay and then usage, that is a challenge." This is a daily calculation and adjustment to ensure that people can navigate each day and is integral to shifting households across different energy networks in operation in these urban spaces.

Energy poverty pushes households into using multiple different interactions with infrastructure and resource flows for cooking. The modern infrastructural ideal would posit that so-called modern cooking services are powered through connections to a centralized network of gas or electricity, as is common in the urban worlds of the north.

In Namuwongo, as in other urban spaces in Uganda, charcoal remains the key energy source for urban dwellers. "Steve" explained, "If they can't use electricity, then they will find other ways that will have implications for the environment." This off-grid cooking infrastructure is widespread. Home cooking with charcoal makes up a remarkably shared practice across social classes in Uganda, especially in the absence of a national gas distribution network in operation, as in cities such as Nairobi. As "Smith", Energy Efficiency Program Officer for UN-Habitat informed me, "We found that less than 2 percent use electricity to cook." Tabitha, a municipal planner in the nearby city of Jinja, reflected on this high usage and future implications, "Interestingly, the middle-income users will still cook with charcoal. It is a very serious issue in whole country, not just informal settlements but almost two-thirds of country, maybe 85–90 percent using charcoal."

The poorest in Namuwongo may not even have the financial capacity to use charcoal regularly. For instance, wood is burnt to make the potent local alcoholic drink and gin alternative, *waragi*. We found out from the brewer that this cheaper fuel source was also supplemented by waste material, including plastics, helping to increase the profit margins for the enterprise, even as the noxious smoke enveloped nearby homes. Charcoal is not the only nonelectric option available for heat and cooking for residents in Namuwongo, again highlighting the heterogenous infrastructure configurations at play across the neighborhood. Dennis at the Ministry of Land, Housing, and Urban Development set out these cooking options, estimating that a "typical person staying in a slum would be living on USD 1 a day. They resort to using firewood, encroach on a forest reserve or use paraffin." "David", a leader of the National Slum Dweller Federation of Uganda, explained bluntly the economic calculations wrapped up in decision making for cooking: "We have people who cook on open fires because of the affordability. A piece of firewood you can get for UGX 500 [¢10] or less, UGX 1,000 [¢20] for three pieces." In effect, these everyday navigations of hybrid and heterogenous infrastructures are survival strategies at work across precarious lives and the techno-environmental geographies of social reproduction.

It is worth considering further the heterogeneity of the infrastructure required to supply charcoal for cooking that Joel and I traced in our documentary *Powering Namuwongo*.[2] Instead of the nonhuman hardware of wires and cables, the substations and the transformers that make up the electricity system, the charcoal infrastructure is noticeably a people-centered

4.2 Workers produce charcoal on an island in Lake Victoria.

4.3 Firewood being used to make *waragi* in Namuwongo.

4.4 A sign communicates the risk of electrification in Namuwongo.

infrastructure. These range from the workers who cut down timber on the islands of Lake Victoria and rural hinterlands to those who chop up the trees and produce the charcoal through slow-burning practices to the boat owners and lorry drivers who transport the timber across the road networks of the country. They also include the wholesale entrepreneurs in the city and the neighborhood traders that work all day and night at their stalls selling to the people, and the entrepreneurs who subsequently use charcoal for generating money from food stands selling products such as the famous Rolex.

The charcoal supply network that we filmed is an extended, off-grid, peopled infrastructure, and it is just one operation amid the heterogenous systems at play in the neighborhood for the supply of cooking fuels. It is replete with unequal and gendered labor relations, dangers, and bad health outcomes for those taking part, even as some make a livelihood and others a nice side hustle. Workers risk arrest to cut down trees on the islands, toiling long hours for little reward. The mainly women entrepreneurs in the city spend all day breathing in the charcoal dust and suffering from respiratory problems. This is the disposability of people acting as infra-structure and facing the "uneven exposure to death" (Doherty 2017, 200) of those reliant on these types of incremental techno-environments. The costs of survival in the city and social reproduction on the margins can often be high. In Namuwongo, as in other popular if resource-poor neigh-borhoods in urban Africa, these hybrid realities are unlikely to change any-time soon. As such, and as the infrastructure involved in the supply and distribution of charcoal impresses, it is imperative to think in more detail about this peopled aspect of incremental techno-environments.

PEOPLED

James Town, or Ga Mashie as it is known locally, is an older neighborhood in central Accra, officially classified as a slum, and with about 125,000 resi-dents (Ghana Statistical Service 2012). The area was a historic site in the development of energy infrastructure in the city. As "Frank," a local politi-cian, told me, "James Town is a historical point of interest for energy in Ghana. It was one of the first places to have electricity, but not for the community until the Nkrumah era." The area now has an aging, creaking, and often disrupted electricity network that is managed by the state-owned Electricity Company of Ghana (ECG). Yet, this formal, centralized infra-structure is facing considerable pressure because of the growing popula-tion, electricity demands, and lack of investment.

Peopled adjustments and reconfigurations of the electricity network are part and parcel of everyday life. During workshops I conducted with Solomon, residents, like in Namuwongo, articulated the difficulty of sus-taining flows of electricity into households, even as they had an official connection to the centralized network. This energy poverty, alongside the lack of state or large-scale market investment into the network to respond, leading to breakdown and malfunction, meant that households became enrolled in the everyday governing of the system, predicated on various forms of social collaboration, calculation, and sometimes contes-tation. Here, Simone's (2013, 243) argument that "people figure them-selves out through figuring arrangements of materials, of designing what is available to them in formats and positions that enable them particular vantage points and ways of doing things" becomes the most persuasive

4.5 Ga Mashie, a historical and popular neighborhood in central Accra.

way to think about the electricity network. These ways of intervening and adjusting mediate the flows of energy into the neighborhood and are always about more than the technical and material.

Clandestine connections are common across households in Ga Mashie as in many poorer neighborhoods across the urban worlds of the south and north. As a very honest "Frank" explained, "Electricity in James Town is a problem. They say it costs too much and they are not able to afford it, so most of us use the illegal connection." Residents and a utility worker at the local ECG payment office suggested that many people are involved in clandestine connections to the electricity network, including ECG electricians operating outside of their official responsibilities. Households undertake these adjustments with the support of electricians who are sympathetic to the needs of households or who are happy to take a small payment. A connection can cost from GHS 50 (USD 6.50).

Another type of modification of the electricity network in Ga Mashie involves households or businesses working again with electricians to configure a "split" electrical supply system. This involves residents registering and paying for a proportion of their electricity through the ECG network and then using a clandestine connection to access further flows of energy. The result is that households appear to be paying for electricity through regular payments of bills, hoping to avoid ECG investigations. This type of subtle alteration is very popular and is commonly used to sustain locally based small enterprises that operate on tight financial margins. Like in Namuwongo, this need for people to become enrolled and engaged in the operations of the network is driven by energy poverty and the pressures of social reproduction and of making a viable life in the city.

Alongside these less than official connections to the network that reconfigure the neighborhood's electricity network in an ad hoc informal basis, other clandestine tactics involving various configurations of people support households in accessing flows of electricity. The introduction of the prepaid meter (PPM) system acted as a mediating technology or barrier to circulations of energy for those households unable to afford the electricity credits from the ECG office, again showing that having a connection to the grid does not mean being able to access this resource flow. PPMs have become a focus of attention for residents who sought to resist the effects of this technology and its purpose to increase revenue for the utility company. These contestations involved two main activities. First, households without PPMs were working together by using the same meter as neighbors and seeking to mislead the ECG about who had responsibility for the bill. However, the ECG caught onto this strategy and was working to stop joint meter usage through introducing even more PPMs. This forced residents to find new strategies of access to the grid. Another collaborative means of connecting underpaid or free electricity was to ask an electrician to come and adjust the meter to stop measuring usage—something that could be done with both the PPM and the older meter. Electricians charged around GHS 10 (USD 1.50) for this work, and social networks helped to disseminate the potential this practice offered.

Tactics and strategies to address energy poverty were not always successful or long lasting. Improvements in accessing electricity could be reversed, could necessitate convincing or bribing a public official, or could result in criminal proceedings against the household or even electrocution and death. Each of these potential outcomes highlighted the disposability of human life and labor that sometimes come with operations of people as infrastructure. The ECG claimed that various types of clandestine activity across the network threatened its revenues, sustainability, and future investment plans. It meant that longer-term costs for the neighborhood might incorporate a persistent state of disrepair and lack of maintenance. The ECG employed technicians to go house to house to find these less than legal connections to the network, while residents attempted to elude detection and forge new connections. Across the neighborhood, as elsewhere, a dialectical dance of disconnection and reconfiguration across the electricity network proceeded that produced an unmappable, ever-shifting infrastructure in flux, in which the roles and actions of various people were integral to its everyday composition.

The multiple collaborations, tactics, and strategies developed by residents in which they became part and parcel of network operations generated an ongoing, low-intensity conflict between community members and the ECG, which played out through interactions of these different actors and the network spaces they sought to direct and control. Because of energy poverty, residents in Ga Mashie had to decide to become enrolled in the network. Their peripheral status in the urban economy and the introduction of PPMs restricting the flows of electricity into the household had a direct infrastructural outcome. In effect, people had to collaborate and form new types of socio-technical arrangement in order to access free or lower-cost electricity. This often came at a cost to the people themselves and always required new strategies when older methods became obsolete as service providers developed new technologies.

Through focusing on the incremental techno-environments of the electricity network in this part of Accra, we see the peopled strategies through which urban dwellers seek to interact, experiment, and intervene with urban energy networks. These incremental ways of reconfiguring the system show remarkable tenacity, despite the actions of operators seeking to ensure that flows of electricity are fully paid for. This is a ceaseless, circulating, and experimental microscale of transformations of urban energy networks that moves between and across the geographies of the city's neighborhoods.

With some level of formal access to electricity, communities such as Ga Mashie are still faced with the need to reconfigure flows through intervening in the network in order to sustain social reproduction because of the complexities and techno-environmental inequalities of contemporary

4.6 Wires in an informal part of Ga Mashie.

4.7 Electricity meters in Ga Mashie.

Accra. And activities, if undertaken for everyday survival, might also open up new possibilities for economic and social advances. Here, the electricity network becomes a site through which urban dwellers might prefigure, imagine, and bring about different types of infrastructure operation. In this way of thinking about people, infrastructure, and incremental techno-environments, we might see a fragile and yet sometime-viable way forward to inhabit the city. Urban dwellers find, consider, test, inhabit, experiment with, and withdraw from particular network constellations dependent on ever-changing circumstances. It is in these urban spaces across and through networks that a significant proportion of residents in Accra live, work, and experience the electricity network. People are always seeking to steer, direct, and open up (incrementally) new possibilities and opportunities through and across infrastructure.

INCREMENTAL POLITICS AND IMPERATIVES

In this chapter, I have shown the need to shift interpretations of the infrastructural beyond the linear narratives of urban modernity as they mutate from the hybrid, heterogenous, and peopled networks that crisscross formal/informal spaces such as Ga Mashie or Namuwongo. These are the lived presents for the many that emerge from the struggles, experiments, adjustments, and interventions across the everyday operations of infrastructure in urban Africa that pose questions about the politics and imperatives of urbanization from below.

Politically, I would suggest, incremental techno-environments vary between a politics of survival and a rather more insurgent and prefigurative disposition. The calculations, demands, and actions that establish such geographies are primarily generated because the state, utilities, or formal private-sector operators will not or cannot provide basic urban services. Alongside this gap in provision, many urban dwellers are living in resource-poor contexts, in which the daily wage for those who can access work remains stubbornly low and the capacity to access or afford connections to formal urban grids if they do exist is also far from guaranteed. Everyday social reproduction is predicated on the capacity to access various resources flows, requiring an array of infrastructural labor, knowledge, struggles, and collaborations. Self-built systems, informal connections, and collective repair and maintenance of these networks are not something people want to do, foremost because this labor takes time away from other activities, but also because these activities are risky and dangerous. The making of incremental techno-environments is driven in part by these everyday imperatives produced through the infrastructural deficit. It is a mode of living and survival that allows people to cope with, navigate, and manage the highly uneven and differentiated ways in infrastructure of the city. This is the politics of infrastructural survival.

Might we also think about these techno-environments beyond survival and rather as a prefigurative politics of infrastructure? This is a way of living the city that is not simply about securing the basic resource flows of social reproduction but also something that is transgressive, sometime clandestine, experimental, and collaborative and which points to the ways interactions with infrastructure can become a critical site of struggle that opens up new possibilities. A prefigurative infrastructural politics is a process of techno-environmental organization and set of social relations that seek to reflect desired futures or an opening of future possibilities. It is both materially and socially prefigured, especially in considering the peopled infrastructures that facilitate these interventions, adjustments, and reworkings of hardware and resource flows. From this standpoint, people generate conditions of possibility through ways of living the city (Loftus 2012) that make explicit the power of urban dwellers to confront various inequalities.

This articulation of the prefigurative rejects an end point in urban struggles around infrastructure. Applied to thinking through incremental techno-environments, these networks are an unfolding set of contested ever-transforming spaces and a series of shifting sites for intervention—a "territorial process where the organizing role played by sovereign institutions and regulations is replaced by sets of fragmented and competing infrastructural scripts, activated by antagonistic agencies" (Minuchin 2021, 5). Here, infrastructure is contained not only in the engineer's domain and the technical plan but as a politicized, hybrid, and heterogenous array of people, technologies, and environments fighting and struggling, collaborating and coexisting, experimenting and sharing.

Prefigurative politics built around infrastructure echoes in Asef Bayat's (2000) "quiet encroachment of the ordinary" and James Holston's (2009) "insurgent citizenship"—claims, forms, and relations in the urban present that place urban dwellers in popular neighborhoods as central to modes of inhabiting, sustaining, and solidifying urban life. For many urban dwellers, clandestine actions are part and parcel of ongoing struggles to survive. They are attempts to secure the essentials of life. This should not prevent a reading of these processes as also a potential emancipatory politics.

If such a prefigurative infrastructure seems at odds with the struggles for universal centralized and homogenous networks operated, maintained, and financed by the state, it might also be both a politics of realism and of a potentially different way of doing things because "whether one is in favor of state-driven development or not, it is self-evident that a large part of the resolution of the African urban crisis will come from the efforts of the urban poor themselves" (Pieterse et al. 2010, 14). The prefigurative infrastructure produced through incremental techno-environments is part and parcel of urban life because the deployment of universally accessible, affordable, and fully functionally central networks can remain a distant promise from the state rather than a likely future. If this scenario is common across large parts of urban Africa, then it poses important questions about whether municipalities and utility companies, development agencies and national governments might support and enable these peopled, hybrid, and heterogeneous infrastructures in ways that empower and support urban dwellers, rather than seeking to close them down, delegitimize, or simply ignore the dangers and risks that come with such operations.

Critical to this political reorientation of infrastructure planning is to recognize the inherent flaws and faults of the logics of the modern infrastructural ideal in guiding the planning and operation of urban networks. As I made clear in the opening chapters, a growing body of work questions this suitability of the modern vision of the networked city across an urban Africa. For instance, Jaglin (2008, 19) has cautioned, "there is also a growing concern, both in policy documents and in the academic literature, that service delivery cannot address the needs of the poor if it relies on traditional technologies and systems that are seen to be inefficient and ecological unsustainable."

These arguments are finding resonance among urban dwellers, municipalities, and some scholars (Nilsson 2016; Lawhon et al. 2018). However, the allocation of significant resources into these networks requires detailed understandings of in situ infrastructures at play in urban Africa. In our

HICCUP project in Uganda with Shuaib Lwasa's Urban Action Lab, the research undertaken by the team showed the potential of these kinds of incremental infrastructure operations across noncentralized networks and infrastructures in the waste and sanitation sectors (Nsangi et al. 2021; Sseviiri et al. 2022). What is required is a political desire to make these infrastructure work, and "from a financing perspective, it is necessary for authorities to invest in integrating, regulating, coordinating, digitizing, and incrementally extending informal and small-scale systems of delivery" (Cirolia 2020, 104).

The experience of urban modernity in the West was predicated on a certain configuration of technologies brought together through the centralized network model. It doesn't mean that urban Africa should follow suit through copy/catch-up. Rather, municipalities and national governments, utilities, and service providers need to take seriously the potential of hybrid, heterogeneous, and peopled infrastructures. This is particularly important in relation to the suitability for decentralized microscale networks and the ways these systems can be extended and potentially upscaled. Recognition that attempts at investments to address the infrastructure deficit through deployment of centralized networks as the only viable future pathway must give way to institutional acknowledgment and fiscal support for heterogenous and hybrid infrastructure (Lawhon et al. 2018) that values, protects, and rewards those people that form the basis for operations. I would contend that claims for universal access to safe and fully functioning infrastructure services should be maintained and indeed enhanced so that these essential services for social reproduction in the city are restricted through neither distributional geographies nor the cost of access for users. How cities deliver these infrastructure remains the critical question.

5

THE EXTENDED TIME/SPACE
OF INFRASTRUCTURE

LIGHTS OUT AND BROKEN TOILETS

In Accra, the lights have gone out again, as have all the electrically run appliances in homes up and down the street as another power cut hits the city. It's not long before the ubiquitous hum of the diesel run generators begins across Cantonments, an affluent suburb and former space of colonial administrators and now postcolonial elites. Lights, air-conditioners, and TVs resume service in homes; bars and restaurants keep serving customers after staff have switched supply to these secondary systems. Social and economic life continues to an extent, despite frustrations over these incessant disruptions in the electricity supply and the associated costs of generators and fuel. Across town, in resource-poor but popular neighborhoods such as Ga Mashie or Old Fadama, people have less access to prohibitively priced generators that cost upward of USD 350. In these areas, urban dwellers are reliant on other forms of light and power, such as candles and torches. Some spaces are noticeably darker and quieter; compounds have closed gates or doors because of concerns with security. When the ECG makes planned interruptions across the network, there is also chatter among residents that these parts of the city receive less supply than the richer areas. Businesses struggle to keep afloat when faced with these disrupted flows. Schoolchildren find it difficult to do their homework for class the next day. People worry about personal safety. They cancel or delay meetings or celebrations. Community life is slowed down. Although everyone in the city is frustrated with the failure of the ECG to keep the grid running, the effects of disruption are not evenly distributed.

In Cape Town, on a hot summer day, a block of communal toilets used by dozens of surrounding households in PJS Section in Khayelitsha are not in a good condition, with many broken. Despite the obligations of the municipality to service these toilets, I can see that maintenance and repair have not been delivered to a sufficient standard to enable people to live dignified infrastructural lives. Some doors are missing. A burst pipe spills wastewater onto the street. Toilets are blocked. The stink makes life challenging for families. Many people in Cape Town have flushing toilets that carry human waste away without a worry or second thought—an invisible and taken for granted infrastructure. In other parts of the city, a different sanitation infrastructure is in operation, which means safe disposal cannot be guaranteed. In these mainly informal, often Black spaces, substandard sanitation services are a reality faced by many communities. This segregated sanitation experience indicates a city deeply divided in the type of technologies, services, and conditions that operate primarily along racialized lines. Nearly three decades into a post-racial democracy, apartheid infrastructure and techno-environmental division are nowhere near being dismantled.

Many neighborhoods in urban Africa remain without safe and reliable infrastructure. To explain these infrastructural injustices means to shift beyond the immediate and adjacent. Understanding the lights out in Accra and the broken toilets in a still-segregated Cape Town requires thinking beyond the technical as the critical explanatory factor. The multiple time/space of shifting power-laden, global–local relations, processes, and operations configure the infrastructure experience. Maria Kaika (2005, 28) described European cities such as London and Athens: "Endowed with modernity's technological networks, the urban fabric became a nexus of entry–exit points for a myriad of interconnected circuits and conduits." To think about modernity's technology networks in African cities is to foreground this experience as an ongoing, unequal spatiotemporal dynamic between metropole and periphery and across the divisions of the contemporary city. Analysis must travel outside city boundaries and into the wider spaces of urbanization, and beyond the immediacy of the present into longer infrastructural and governance histories. Doing so makes visible the logics, power geometries, and inequalities that inscribe contemporary operations.

DECENTERING THE CONTEMPORARY CITY

Decentering the time/space of the contemporary city in the Infrastructural South shifts analysis of "urban" spaces beyond city boundaries, and time outside the present moment as constitutive of the making of infrastructure. Doing so is vital to challenge and call out explanations that depoliticize infrastructural injustices and ascribe issues such as electrical disruption or unsanitary toilets as purely technical problems to be solved through techni-cal means. I introduce the *imposition* techno-environment to draw attention to the making of technologies and environments as a form of extended urbanization and broader constellations of political and economic power and governing.

To think about the time of urban infrastructure is to think about "the myriad ways in which infrastructures are constantly transforming as they erode, degrade, and are devalued, updated, and maintained" (Addie 2022, 110). But it is also to consider how shifting political ideologies, histories, and logics imbue regimes of infrastructural governance into the future (Moss 2020). To think about the space of urban infrastructure is to consider the "global socio-natural dimensions of urbanization that span city and coun-tryside" (Angelo and Wachsmuth 2015, 20). This is to emphasize how meta-bolic flows that produce techno-environments draw in wider hinterlands of operation—a fundamental anchoring idea within UPE (Swyngedouw 2004). This approach incorporates both a multi-scalar and relational analysis of infrastructure that encompasses cities, metropolitan regions, and extended landscapes of extraction, production, logistics, and waste-management.

Intersections and constellations of network histories and extended geo-graphical, metabolic processes span across existing configurations, opera-tions, and experiences of urban networks that Omar Jabary Salamanca and I (2023) describe as "inscriptions" imposing racialized political economic regimes and modes of governing infrastructure into the present. Inscrip-tion brings to the fore how unequal, center–periphery relations remain paramount to today's geographies of infrastructure and draws on the work of scholars such as Ann Stoler (2016, 33) who attempt to think "between a past that is imagined to be over but persists, reactivates, and recurs in trans-figured forms." This extended time/space of urbanization has been a critical part of historically grounded studies of urban networks in the South (Kooy and Bakker 2008; McFarlane 2008) demonstrating how "colonial sensibili-ties, distinctions, and discriminations are not just leftovers, reappointed to other time and place" Stoler (2016, 33). We should not restrict focus on infrastructure inheritances in African cities to history, but rather we should see them as overlapping and active presents and, as Kimari and Ernstson (2020, 3) suggest, a "scaffold" to the contemporary, unequal experiences of infrastructure.

THE EXTENDED TIME/SPACE OF DISRUPTION

In 2015, Accra was experiencing serious disruptions of electric supply, with these power cuts termed after the Twi word *dumsor*, meaning on/off, and which continue to this day. People were going eighteen hours a day with-out electrical power. These interruptions were costing the nation around 2.6 percent of its gross domestic product (Eshun and Amoako-Tuffour 2016). There was widespread anger, seen in street protests, alongside frus-trations from households unable to go about everyday life. As "Bell," a Ga Mashie resident explained during a previous trip, "We depend on light for our everyday activities. Without light or energy, there were many prob-lems for people trying to make money or for people to get by in the family compound." For many in the city, the promise of infrastructure was not only fraying but also under pressure of collapse. "Bob", who worked at the ECG payment office in Ga Mashie explained, "The lights out affect a lot of people and business, with much revenue lost during these episodes, disturbing people's lives and ruining appliances and business." This lack of power in the city posed the question as to why Accra was struggling to provide a reliable supply of electricity, much like in Soyinka's novel (1965, 86) *The Interpreters* set in 1960s Lagos in which "the street lamps were just beginning to flicker in their eternal struggle against uncertain power. It would continue for half the night and then perhaps the duty engineer would find the faulty coil and take it out altogether, leaving the street in darkness for a month or more." Analysis of this disruption in Accra, as it would in other cities such as Lagos, shows the story of urban electricity in the city as an extended one. It brings into view the overlapping eras of governing of the national grid that draw in places far away from the streets of Ghana's capital as integral to the production of *dumsor* and contemporary, lived, and unequal experiences of disrupted infrastructure.

COLONIAL POWER

Colonial-era infrastructure in Accra had much in common with those in other urban areas across the region. It emerged within the geographies of imperial operations and the administrative apparatus required to manage various resources, including precious metals from the Ashanti region (Quarcoopome 1993). The transformation of natures into flows of capital underpin urbanization (Cronon 1991; Swyngedouw 2004). Colonizers extracted these resources and partly reinvested them in the city, leading to further growth, as well as being sent back to finance the early modern networks of the metropole.

The material history of electricity in Accra is of a racialized, segregated system justified through urban planning predicated on discourses of racial supremacy. This history began with the 1896 Town Council Ordinance introducing streetlights (paraffin lights). Colonial authorities subsequently established committees to develop urban service provisions for a range of services, including lighting (Dickson 1969). From 1914, they generated limited electricity for colonial industries in nearby Sekondi, and by 1922, a newly established Public Works Department provided electricity to fulfil the needs of the imperial industries and the residencies of the colonial elite in Accra. By the early twentieth century, spatial segregation, mandated through law, established divisions not only of people but also of urban services such as electricity. As Graham and Marvin (2001, 82) suggested, in colonial space, "this partial completion of modern infrastructure was a very deliberate attempt to symbolize the superiority of colonial power holders over colonized civilizations."

In 1947, colonizers established the Electricity Department in the Ministry of Works and Housing—a dedicated unit to oversee the growing but segregated electricity network in Accra. Following the Second World War, as mobilizations for independence gathered pace, the colonial authorities moved toward a somewhat paternalistic rule as discourses of modernizing Africans came to the fore (Iliffe 1987). This shift in infrastructure governance showed a changing emphasis by the colonial authorities to begin to extend electricity services outside colonial enclaves.

POSTINDEPENDENCE MODERNIZATION

Ghana became independent in 1957. Its first president, Kwame Nkrumah, set about a program of modernization built out of an Afro-socialist vision of the future (F. Demissie 2007). The architecture of the era represents this rupture from the old world, particularly in the government buildings and the epic space of Independence Square. And it wasn't just in the architecture that these visions found material expression. Electricity formed a key pillar of Nkrumah's postindependence program of modernization. He argued (1965, 248), "We are convinced that the accelerated development of the country will depend primarily on our success in developing cheap power resources."

In 1961, the Ghanaian government established the Volta River Authority (VRA) with USD 50 million of investment focused on the planned generation of hydropower. The ECG followed this utility company in 1963, responsible for supply and distribution. The modernization era was a break from the logics of colonial rule. Yet, it continued to be mediated by ways of governing the city and the (fragmented) legacies of previous eras of infrastructural governance. This was especially the case in how the postcolonial elite took over former colonial and electrified spaces such as Victoriaborg for themselves.

In the city, after massive investment by the government in power generation, increasing numbers of neighborhoods were linked to the electricity network in subsequent years. Memories of this electrification remain for some in the city. "Kojo," an elderly resident of Ga Mashie, told me, "When Nkrumah was president, he made promises to the nation that we would have the electrical power and he constructed the dam." "Komla," a neighbor of "Kojo," remembered the day that his household first got connected and recalled, "I felt the power of it." And "Enam," a local politician with the Convention People's Party, recalled he "was told in infancy they used to use two stones to create light and then one day we could press a switch and receive light."

Akosombo Dam and its power station stand testament to the modernization ideals of the Nkrumah government, which sought to universalize electricity supply in the cities and villages through generation of hydroelectric power. Nkrumah spoke of such importance for the post-independent era in which, "newer nations, like our own, which are determined to catch up, must have a plentiful supply of electricity" (BBC cited in New African, 2018, n.d.). His vision required a feat of geological and technological engineering on a scale rarely seen anywhere on the planet. Ghana constructed the largest man-made lake in the world (at the time), covering more than 8,000 km² (Fobil and Attuquayefio 2003), with a series of often disastrous social and environmental impacts. These included the displacement of more than 80,000 people in the Volta River basin (Gyau-Boakye 2001), many of whom have been resettled in villages close by, where they continue to suffer from health, social, and economic problems. Indeed, during a visit in 2015 to one particular village, I was shocked to find out authorities had still not electrified the settlement all these decades later.

5.1 The modernist architecture in Independence Square.

5.2 Akosombo Hydroelectric Plant.

The construction of the USD 324 million Akosombo Dam made up of loans from the UK and US provided a massive financial commitment that enabled the distribution of electricity to a growing number of Ghanaians as it transformed flows of water from the Volta River basin into an urban service provision.

The colonial authorities conceived plans for the Volta River project from the 1920s, with a feasibility study conducted by British engineer William Halcrow in the late 1940s. The project's colonial origins were inscribed into the future, even if they were never to be implemented because they acted as a "scaffold" (Kimari and Ernstson 2020). This is because they established a technological solution that was taken up by the post-independent government. Despite the end of the colonial era, the dominance of the World Bank in Ghana's developing energy policy was clear, primarily because this newly independent African nation-state had limited access to finance. To power the nation, the government needed to secure international investment and technological expertise. This entailed forming an agreement with the American-based Kaiser Aluminum and Chemical Company, creating, with World Bank backing, the Volta Aluminium Company (VALCO).

The conditions of the financing arrangement, to which the Ghanaian government contributed around 50 percent, meant that the priority of the VRA at Akosombo was to power the VALCO aluminium smelter and at a discounted price, and only then distribute to the rest of Ghana. Such arrangements reflected the difficulty newly independent states experienced in generating finance for infrastructure investment. The massive costs of this project prevented government investment in building other power-generation facilities, tying Ghana to this one energy generation space—a decision that would reverberate down the decades. Following multiple, overlapping financial crises that culminated in 2006, VALCO became state owned, but the company remained the largest consumer of power in the country, estimated by a VRA representative I spoke to at up to 40 percent, for which it pays 25 percent less than other consumers.

Akosombo Dam and indeed the adjacent Akosombo Township, the modern town built at the same time, remain compelling articulations of an era of Afro-socialist visions of modernization. Electricity became the symbol through which Nkrumah could establish a pathway to a future in which Africans might use Africa's nature, rather than imperialism exploiting it. The government had transformed water into electricity. As "Joyce," a VRA worker at Akosombo Dam, told me, "In Ghana water is life, not just for the thirsty but for those who need energy in their lives." It also brought public expectations that the government would provide essential services to all villages, towns, and cities. Underneath this vision, contradictions and the inherent difficulties of these modernization programs in post-independent Africa surfaced because of a reliance on foreign finance and political interference related to the Cold War that meant the World Bank and its US and UK backers had significant influence over Ghana's energy policies. Despite Nkrumah's (1965) recognition of neocolonialism as the biggest threat facing post-independent Africa, he was to fall victim to a US-backed military coup in 1966. The consequences of the coup brought about a marked decline of modernization visions and associate investments for the country (Fitch and Oppenheimer 1966).

STAGNATION AND ADJUSTMENT

During the era of alternating civilian and military rule (1966–1992), the electricity system in Accra continued to expand across the city. However, attention on energy generation to keep up with demand declined markedly, despite what Brenda Chalfin (2010, 197) termed "the later copycat projects of Acheampong" and after the final completion of Akosombo in 1972. This failure of the government and of its underfinanced utilities in this period to develop a strategic response to the electricity generation needs of the country, including through diversification beyond Akosombo, is a key historical factor in the disruption of electricity in Accra. The need for electricity from households compounded demand, as domestic consumption doubled between 1967 and 1976 (Resource Center for Energy Economics and Regulation 2005). As Winfred Nelson at the National Development Planning Commission told me, failing to keep up with demand was now pressing and "the ECG has calculated 500 people should be reliant on a transformer, but this is nearer to 1,000 because of the urbanization, so there is a lot of pressure."

Neoliberal reforms from the 1980s intensified underinvestment in the electricity network brought about, at least in part, by the huge debt that came with the loans to construct Akosombo Dam, with repayments only ending in 1987. Chalfin (2010, 195) described Ghana as being at the "forefront of neoliberal reform in Africa." Structural adjustment left the electricity network in Accra increasingly vulnerable to disruption because of the fiscal constraints faced by the utilities and subsequent failure to invest, maintain, or repair. During this period, structural adjustment had significant consequences for electricity generation capacity. The Power Sector Reform Programme, instigated from the mid-1990s by the World Bank, entrenched neoliberal logics within the electricity sector as it sought to shift financing from government and donor sources toward the private sector by encouraging new independent power producers to enter the market. Energy sector liberalization, ongoing since the 1980s, laid the groundwork for the government to sign the Ghana Power Compact (GPC) with the US government in 2014.

The GPC proposed a radical transformation away from the state control that had characterized the sector's history. They framed it as a "solution" to address ongoing underinvestment in the electricity network and the multiple operational issues faced by the utilities, including high transmission loss, an obsolete infrastructure, and poor management. Subsequent part privatization of the ECG followed through a concession for a New Zealand/Philippines consortium to manage the utility for twenty years from 2019 built on the logics of market-based solutions to Ghana's energy woes. However, less than a year afterward, the government terminated the contract because the new concession "could not solve a key issue underlying the privatization of the ECG: dum sor" (Ohemeng and Zaato 2021, 5). The market was not, in fact, the solution to these disruptions.

EXTENDING THE GEOGRAPHIES OF DISRUPTION

Analysis of the production of *dumsor* in Accra requires thinking through the overlapping eras of governance that draw attention to the time of infrastructure. Within these histories, we can also see the more than urban geographies of the contemporary experience of disrupted electricity in the city. Critical to such an extended space of infrastructure is the role of Akosombo, nearly 100 km from the city, and its global making through histories and relations with the UK and the USA. This hydroelectric power complex also illustrates another key space beyond the boundaries of the city that are integral to the disruption of electricity in Accra: its hydrological zones that provide the flows of water necessary for electricity production in Accra stretch north toward the Sahel. This is because of the growing impact of climate change on the capacity to generate power at the dam. These catchment areas are experiencing increasingly unstable rainfall, which is in part attributed to the effects of climate change in this fragile region (Kandji, Verchot, and Mackensen 2006). The Sahel areas of northern Ghana, and the other West African countries that make up the Volta River basin, are already experiencing reductions in rainfall of 20 percent over the last twenty years (CARE 2007). This vast geographical zone is becoming increasingly arid (Gyau-Boakye 2001), which may further reduce rainfall by up to 27 percent (CARE 2007). Processes of deforestation and desertification in northern Ghana, closely implicated in the urbanization process and the energy needs of urban dwellers through charcoal, have further reinforced the growing aridity. The results are dramatic. Forecasts show the water supply for Akosombo Dam will experience reductions in flow of

5.3 Electricity infrastructure in Ghana.

between 30 and 40 percent over future decades (EPA 2000, 6), leading to a reduction in hydropower output of up to 59 percent. This is the extended space of infrastructure at play in contemporary operations that further threatens the capacity to keep electricity flowing to the city. Winfred was worried. He told me, "We have all been terrified by the fact that it [the water levels] gets to some points when its reported in the newspapers that it's at a low level."

THE EXTENDED TIME/SPACE OF UNEQUAL SANITATION

They were forcing people to smell their own shit.
—Curate Mawethu Ncaca

Infrastructure has been a tool of planning and a lived experience of control across and beyond urban space that reflects and reinforces historic and contemporary forms of racialized political economy (Nemser 2017; Jabary Salamanca and Silver 2022). This extended time/space of infrastructure remains ever present in Cape Town as an ongoing imposition of colonial and apartheid urban planning and its inscriptions into the postapartheid era. Cape Town makes visible the racialized operation of sanitation infrastructure and how it orders technologies, flows of human waste across the city.

The scene may have come straight from the lens of photographers such as Santu Mofokeng or George Hallett, whose harrowing work documented segregated urban life under the apartheid regime. In June 2013, Mayor Patricia de Lille of the Democratic Alliance (DA)[1] visited the informal settlement of "Barcelona" in the township of Gugulethu, which is about 15 km from the center. She arrived with a large police escort and accompanying armoured "Rhino" vehicles that have come to symbolize state violence against poor Black communities in the so-called Rainbow Nation. The visit took place as the settlement had experienced the breakdown of the maintenance regime of its toilets because of strikes by workers. The mayor wore a surgical mask as she entered Barcelona, deeply upsetting the residents. To make matters worse, she only stayed briefly, perhaps finding the stench that residents were being forced to live in too much. There was little time to talk to residents facing unsanitary conditions. The mayoral visit was merely another act in what quickly became a highly contentious moment in the city in which sanitation injustices had spun into something larger than the operating of infrastructure and shone a spotlight on the long-standing, pervasive racism that segregated a city in one of the planet's most unequal societies.[2]

In 2014, I visited "Tim", a street committee leader in Barcelona, to find out more about the visit of the mayor and the sanitation issues faced by residents. We entered his home as two security men stood outside. "Tim" was a powerful figure in the community, judging by his protection. Reflecting on de Lille's visit the previous year, he said, "If she can't stand five minutes, then why do we have to stand it for three months [waiting for buckets to be collected]?" It was a pertinent question that captured not just the contemporary moment but the extended time/space of infrastructure in the city. In Barcelona in 2016, each of the 117 toilets in the Barcelona settlement was shared by twenty-three households—much higher than the city standards of one toilet to five households (Western Cape Government 2016). It classified most of these toilets as buckets. The bucket system is a stark reminder of how apartheid lives on through infrastructure across South African cities. The bucket system was initially called the Rochdale System of sanitation and was used in Victorian Britain across the rapidly urbanizing, under-serviced conditions many industrial workers lived in. That it was still being used in South Africa was a source of great anger for many, especially considering the immense wealth of parts of the population in cities such as Cape Town.

"Tim" explained the bucket system as the most basic type of toilet technology, which was often located inside or near a household, sometimes with a structure built around. Community members used the bucket to collect faeces, which was subsequently picked up by a service provider and emptied. This basic sanitation infrastructure has had many downsides for the community, including the lack of dignity it affords households, the odious smell, and the health risks. "Tim" described in vivid detail the conditions that residents faced in the settlement: "There are maggots in the buckets after two days. It is very dangerous. When the toilets were not picked up, health professionals came, as lots of children [were] going to hospital." He told us of the frustrations of the urban poor in the new South Africa that had hoped for something better: "We have the bucket system here, and people don't like it. President Mbeki promised that there would be no buckets in the country in 2014, but look now. We have tried for three years to push strongly against buckets." It was reported that across South African municipalities, more than 60,000 households were still using the bucket system as of 2017 (Statistics South Africa 2018), with the figures actually increasing in the Western Cape Province (the region in which Cape Town is situated) by more than 10 percent.

The bucket system requires regular emptying in order to limit its health impacts on households. In Cape Town, it was the municipality's responsibility to ensure regular servicing of this technology, and it was the breakdown of this maintenance regime that led to a political flare up in Barcelona. In 2011, and bowing to public pressure as she started her new role, de Lille funded maintenance services for the sanitation infrastructures operating across the informal settlements of the city. Using finance from the Expanded Public Works Programme (EPWP), a national fund, rather than the local budget of the city, the municipality invested up to R138 million (USD 9 million) in a new janitorial service. The aim was to ensure

5.4   A broken "bucket" toilet in Barcelona informal settlement.

maintenance of the various toilet technologies in operation in Cape Town. This investment was from the start underfunded, and this meant that the private operator, Sanicare, was forced to cut the hours and pay of workers, predicating a labor strike in 2013. The strike meant buckets were not being collected and were left in situ for up to three months. It is these conditions that led to exasperated residents wondering how the municipality could treat them in such a way. "Tim" explained, "We tried to call the City of Cape Town about strikes, no one [was] interested," and later after the visit, "She promised to come back. This is when we got really angry."

Tony, like other people I spoke to, suggested that the reason for this lack of a proper maintenance and repair regime and the valuing of infrastructural labor required to make it work in Barcelona was because "they are using the fact that people are from Eastern Cape—this is political drive." He made clear the racist attitudes that are at play in some parts of Cape Town concerning the migration of mainly Black South Africans from the poorer province to this city.[3] The toxic notion that so-called refugees from the Eastern Cape are stopping Cape Town from being able to deliver for its local population has gained traction among some white people. These racist tropes of Black South Africans as somehow unmodern and outsiders and therefore undeserving of decent infrastructure seemed to have entered into both some parts of the white population and the highest level of political life in the city. It enabled these disinvested regimes of maintenance that should have been abolished in the 1990s to continue.

Given the breakdown of maintenance regimes, things spiraled further in Barcelona as residents' anger about their conditions created a desperation to get the municipality to fix the problem. It resulted in the Cape Town "poo protests" (Robins 2014). A newly formed group of political activists closely associated with the African National Congress (ANC) led the protests, the Ses'khona Peoples Movement. These activists and their actions suddenly if briefly burst onto the scene in the city and, through their actions, made visible the extended time/space of infrastructure. In speaking to Sithembele, a Ses'khona spokesperson, one day in Khayelitsha, he told me about how the frustrated Barcelona community had called in his group to resolve the situation. He explained, "They decided to take the decision, [and] they dumped the poo on the N2. People were happy dumping it. People lifting buckets and staged protest; it was chaotic. It had an impact that cannot be seen again."

After the distribution to one of the main highways in the city, the community and Ses'khona changed their target of anger toward those implicated in this racialized infrastructural inequality. Sithembele described what happened first at City Hall in which they "decided to put the poo at the doorstep, to smell it, how it smells in a bucket in Barcelona" and then "to go to airport because it is one of the things that government is grandstanding about." The aim was to disrupt the centers of political and economic power in the city. The reaction of one arrival at the airport perhaps best symbolized the shock of these acts by telling a reporter of the Cape Argue, "It's not supposed to happen. I have never seen anything like that anywhere in the world."

The poo protests made clear to the public in Cape Town, as "Tim" had done differently in Barcelona, how the apartheid ordering of infrastructure reverberated into the contemporary moment across an unequal city. Until 2015, the slogan of the City of Cape Town was "This City Works for You," but as residents in Barcelona knew full well, this promise did not extend

5.5 Ses'khona activists gather in central Cape Town.

to households struggling to access sanitation infrastructure in the poorest (and Blackest) parts of the city. The City of Cape Town defended the hollow slogan, despite the damning evidence of their lack of care for their communities. Tammy Carter at the South African Human Rights Commission made clear to me this disparity between the slogan and the reality and the priorities of the municipality. She explained, "Last year before the elections [someone] sprayed on toilets 'This City Works for a Few.' . . . [It was] amazing how quick it was eradicated compared to fixing the toilets." The rapid response to defending the city brand was indeed indicative of a municipality that seemed unwilling to provide decent sanitation and accompanying maintenance regimes for Black residents, even as it worried about its image for tourists and investors.

IMPOSING POLITICS

In this chapter, I have attempted to convey the making of the Infrastructural South beyond the immediate and adjacent. To do so, I have proposed the imposition techno-environment. This is an extended time/space of infrastructure that configures the contemporary experience of urban life in cities such as Accra and Cape Town. We cannot only focus on technicalities when investigating failures in the operation of infrastructure, whether disrupted electricity flows or substandard sanitation provision. Analysis must shift beyond the fault that can be fixed or a challenge that can be overcome through more development, technical expertise, or innovation. Doing so makes visible the lively histories and geographies that produce today's infrastructure conditions.

In Accra, my analysis of disrupted electricity in the city offered one way to think about the extended time/space of infrastructure. The inscriptions of the past regimes of governance produced disruption as an active process in the present. Kwame Nkrumah understood how the political economies and power geometries of colonialism continued after independence. To think about the contemporary electricity network meant focusing on the ways colonial administrators shaped a segregated electricity provision that prioritized elite spaces in the city and, in the opinion of some poorer residents today, still demarcates who gets the best service or the least disruption. Nkrumah went further though than simply implicating the past as what we might term an infrastructural inheritance. He argued, "Investment, under neo-colonialism, increases, rather than decreases, the gap between the rich and the poor countries of the world" (1965). The intent of Nkrumah was to draw attention to the perpetuation of center–periphery relations in which the end of colonialism did not open a future beyond imperial control and underdevelopment. Rather, it began a new phase of intervention, reinforced by Cold War calculus in which the metropole continued to assert its power and priviledge to shape regional and global futures. The implications for understanding contemporary disruption are that the Nkrumah government, guided in part by previous colonial-era plans to construct Akosombo, was reliant on financing from the West to build such as massive power generation facility. This not only created a dependence on hydropower that would have later implications, but also prioritized relations with VALCO at the cost of the wider nation.

Direct attempts at what Nkrumah termed "neocolonial control" would give way to more indirect forms of intervention as the structural adjustment program (SAP) rolled out across Ghana. A neocolonial reading of the

SAP would suggest this was an intervention by the West to push Africa away from Afro-socialist visions of post-independent states into the marketized and commodified world. It had disastrous consequences because "The Bretton Woods institutions are like arsonists, lighting new social fires" (Toussaint 1999, 30). In Ghana, this hindered the Ghanaian government's attempts to diversify the energy-generation mix and address the growing obsolescence of the network (within and beyond Accra), despite rapid growth in demand. The long-lasting impact of structural adjustment on public investment then created conditions through which Ghana traveled down a path of liberalization with privatization billed as a solution to long-standing problems—a fix that soon failed.

This extended time of infrastructure was also an extended space that connected growing climate change impacts, historic and global shifts in circulations of carbon, and resultant hydrological dynamics and drawing in various zones from northern Ghana and the wider Sahel region to those countries in the north most responsible for greenhouse gas emissions. It incorporated the global relations in which a newly independent Ghana was situated, including the Cold War, and the historic geographies of colonialism in which decisions about urban infrastructure, investments, plans, and futures were made in the Colonial Office in London as much as Accra. The contemporary urban infrastructural is produced through and by these global relations that extend far beyond the boundaries of a city.

In Cape Town, the reverberations and inscriptions of colonial and apartheid governing infrastructure lurch and mutate into the present, and can help explain contemporary sanitation geographies and experiences of residents in Barcelona. Drawing on the longer histories of how space in Cape Town was ordered in colonial and then apartheid eras through racist ideology and racialized urban planning highlights what Maynard Swanson (1977) termed the "sanitation syndrome." This is an inscription that arguably shapes the present in powerful ways. Explanations of contemporary sanitation should draw on a history in which the racialized construction of Black bodies as contaminated and diseased gave credence to the need to segregate urban populations. For instance, after an outbreak arrived from Argentina in the early twentieth century, the colonial authority enacted a Public Health Act to remove 6,000–7,000 Black people to the planned township of Uitvlugt/Ndabeni. And this inscription continued into the apartheid era as the government used panics about health and sanitation to forcefully moved non-white residents from central areas such as District Six to areas in which "there was systematic under-investment in municipal infrastructure" (South African Department of Provisional and Local Government 1998, para. 2.3).

Returning to Fanon's *The Wretched of the Earth*, we can see how the settler-colonist logic operated because it "dehumanizes the native, or to speak plainly, it turns him into an animal. In fact, the terms the settler uses when he mentions the native are zoological terms. He speaks of the yellow man's reptilian motions, of the stink of the native quarter, of breeding swarms, of foulness, of spawn, of gesticulation" (1967, 42). Through such racialized discourse, Fanon showed how the racist logics of colonial planning shaped cites in which "The settlers' town is a strongly built town, all made of stone and steel. It is a brightly lit town; the streets are covered with asphalt, and the garbage cans swallow all the leavings, unseen, unknown and hardly thought about" (38). This compares to the "The native town is a hungry town, starved of bread, of meat, of shoes, of coal, of light. The native town is a crouching village, a town on its knees, a town wallowing in the mire." Fanon would have known firsthand in his youth in Fort-de-France, Martinique, about such segregated conditions, as David Macey (2012, 52), in his biography of Fanon, makes clear, arguing, "Sanitation was not good and the open drains—home to thriving populations of rats and land crabs—was not a pretty sight."

These unequal, contemporary experiences of sanitation are, as Steve Robins (2014, 104) argued, tied to "deeply historical process of racial capitalism" that we can trace back to earlier inscriptions of colonial/settler colonial governing of infrastructure. This imposition techno-environment shapes the postapartheid city in powerful ways because racist governing of infrastructure is actively reproduced across time/space. The material legacies of colonialism and apartheid were not consigned to the past. Inscriptions of an extended time/space of infrastructure didn't fall when the government released Mandela, nor when the ANC got into power, nor twenty years after the formal apartheid system ended. These inscriptions and the imposition techno-environments they produce remain painfully visible and seemingly actively sustained in neighborhoods such as Barcelona and across the city more widely. In effect, the local and provincial

5.6 Toilets in Khayelitsha.

states have not fully dismantled the apartheid that structures everyday experiences of infrastructure in Cape Town.

In Accra and Cape Town, I have shown in this chapter how the imposition techno-environment is a force that draws in the extended time/space of infrastructure. I explored how it operates and is experienced in the contemporary moment but is shaped and suffused beyond the immediate and the adjacent. Most of all, I would contend, the imposition techno-environment makes visible how the Infrastructural South is produced in a myriad of extended, unequal ways that reinforce and reproduce longer histories and wider geographies of racial capitalism.

6

PROMISES OF DEVELOPMENT, EXPERIENCES OF DISPLACEMENT

REHABILITATION

I had entered the fenced-off Port Bell in late 2018, which had had its security increased since a visit a few years earlier. The water lapping the shores of Lake Victoria remained covered in a thick layer of hyacinth, and the landing area for local fishermen and charcoal from nearby islands, adjacent but outside the port zone, remained animated with all types of hustle, despite recent development. The Central Corridor connecting this part of Uganda to the Indian Ocean port city of Dar es Salaam was now, in part operational. The series of investments into the rehabilitation of this lakeside port were influencing the speed and intensity of circulations of various resources, goods, and products between hinterlands and export markets.

Our guide around the facility, "Phillip," had started employment in 1977. This was the period when the government of Belgium made the last set of improvements, constructing a dry dock aimed toward expanding natural resource extractions from the borders of the DRC more than 500 km away. "Phillip" told us of the years of service when nothing much moved at Port Bell as its cargo loads dropped from 600,000 to 10,000 tonnes annually (Muwanga 2018). He was pleased to see Port Bell animated again. In many regards, the underlying resource flows have stayed the same over the years as when the dry dock was built four decades previously and perhaps even stretching back to its colonial inception.

The first boat we came upon was exporting timber sourced from eastern DRC and arriving via Lake Albert, destined for the international markets through the Tanzanian lakeside port city of Mwanza and onward to the Indian Ocean. It was being loaded up to a Kenyan-registered ship whose captain explained he was still learning how to navigate Lake Victoria after being shifted from years of service on ocean-going vessels. Meanwhile, Uganda Railways Corporation carriages loaded with iron sheets were waiting nearby to be rolled on to another ferry. They were being exported from the Baati factory established in 1964 in the nearby industrial zone, one of two in Uganda and part of the Safal group, Africa's largest steel roofing company in Africa.

The rehabilitated railway line, which connects Port Bell, Kampala, and the Ugandan hinterlands to the Indian Ocean, went right up to the largest ship in the dock, which itself had standard-gauge track to allow the goods to be deposited directly onboard. The 2,000-km road journey to Dar (or Mombasa along the competing Northern Corridor) was now in competition with a multimodal rail/boat/road trip that offered the promise of increased capacity, competitive pricing, and ultimately enhanced national economic development. As David Wangi of the Kampala City Traders Association made clear in the *Monitor* newspaper, "What we need from you is service which is cheaper, faster and efficient. Then our job will be made easy in Kampala" (Nakaweesi 2017). Integrating Port Bell and the connecting railway into the Central Corridor infrastructure project started by Tanzania had seemingly addressed the needs of traders in Kampala.

The promise of tens of billions of dollars of investment flowing into large-scale infrastructure projects such as the Central Corridor seems poised to transform urban spaces, as has slowly happened at Port Bell. Corridors sometime take the form of investments in basic transportation systems, such as road improvements, especially into extractive, resource hinterlands such as western Uganda. More often, however, planners configure the projects as multimodal transportation and economic visions, incorporating upgraded and new ports, logistical hubs and special economic zones, multiple rail and road investments, and accompanying real estate developments, particularly the enclaves and new cities examined in chapter 3. The

African Development Bank Group is clear in the purpose of these large-scale projects "as a tool for stimulating social and economic development in the areas surrounding the route" (Mulenga 2013, 1).

In this chapter, I ask what are the implications for cities and understandings of the Infrastructural South of the surging numbers of corridors being planned, financed, and operationalized across sub-Saharan Africa? The in-the-making techno-environment of the corridor generates more than urban geographies produced through expanding regional and global networks of trade, intensifying extractions, and shifting financial calculations and built environment transformations. Again, as in the previous chapter, this techno-environment problematizes a focus on the contemporary city as the only time/space through which to understand the urbanization process. This is because the deployment of corridors extends the idea of the urban beyond municipal and metropolitan boundaries. Governments and other actors design these large infrastructure projects for transnational, logistical circulations and flows of people, goods and finance, ideas and exchanges, and because they build on historical legacies, inscriptions, and echoes of previous initiatives that connected regions into global geographies of trade and extraction. If corridors are being deployed as speculative planning technologies and are full of assurances of growth and development, this does not mean such infrastructural promises are always being actualized. Rather, the techno-environments being generated vis-à-vis the corridor pose uncertain futures for national economies and, as I will also demonstrate, everyday urban life.

WHAT IS A CORRIDOR?

The intention of planners and politicians in deploying corridors is to use infrastructure to organize national and regional territory and space around the logistical operations of global capitalism (Schindler and Kanai 2021). The consequence of these profound transformations is the way the world is ordered. As Keller Easterling (2014, 15) noted, "some of the most radical changes to the globalizing world are being written, not in the language of law and diplomacy, but in these spatial, infrastructural technologies." Cities emerge as intense sites of these logistical networks through the various types of zoned space and volumetric, built environment transformation which are critical to the broader operations and geographies of extended corridors. This is because "despite their rhetorical depiction as continuous surfaces, the spatial configuration of the flows of capital, goods and people are expected to predominantly occur across a network of urban nodes" (Apostolopoulou 2021, 832).

The ever growing techno-environments of the corridor demand new understandings of how finance pouring into the making of these projects might reinforce or address existing urban inequalities and power geometries, as much as the capacity of these investments to enable new types, forms, and intensities of circulation. To do so pushes analysis to consider two related processes: first, the massive, speculative investments, plans and operations of corridors; and second, how these transformations reconfigure contingent, fragile social infrastructures already in operation in urban space, now demarcated and zoned for such projects. It is at these scalar intersections of the massive and the micro that we can gain a sense of the techno-environments of the corridor as integral in the shaping of Africa's urbanization—a delicate dance between promises of development and experiences of displacement.

Longer trade histories throughout the region that leave "inscriptions" anchor the making of contemporary corridors. This is because "the spatialities of the corridor and the ways these are *inscribed* in multiple and overlapping temporalities" as I write about with Omar Jabary Salamanca. These incorporate the precolonial routes, such as the Trans-Saharan (Lang 2017) originating in the eighth century and traveled by Mansa Musa's legendary caravan of people and gold in the fourteenth century from Timbuktu to Mecca (Abbou 2016), and in East Africa, the Indian Ocean–based trading traditions of the Swahili coast (Nicholls 1971) and medieval-era ports such as Kilwa Kisiwani. The infrastructure of colonial extractions imposed through ever-intensifying, often speculative railway technologies into resource hinterlands (Gwaindepi 2019; Jedwab, Kerby, and Moradi 2017), such as the Dakar–Niger line featured in Ousmane's (1995) *God's Bits of Wood* and the postindependence, Pan-Africanist visions of infrastructural integration in the 1960s and 1970s (Cupers and Meier 2020) explored in Thiong'o's (1977) novel *Petals of Blood* and seen in projects such as the TAZARA railway line out of the Zambian Copperbelt (Monson 2009).

6.1 The Tazara railway, connecting Zambia and Tanzania.

It is worth briefly considering the implications of these inscriptions for contemporary analysis of large-scale infrastructure projects, first, to remind ourselves of the importance of precolonial civilizations in the region in the shaping of the world economy through early forms of globalization and modernity (Abu-Lughod 1989; Green 2019), and second, to highlight how postindependence leaders were faced with a range of poisoned colonial legacies pertaining to the infrastructures of trade, movement, and exchange (Chimee 2020), or what Ndlovu-Gatsheni and Mhlanga (2013, 13) termed the "bondage of boundaries." Here, I draw attention again to the making of modernity in the metropole as relational and wholly reliant upon the infrastructures of colonial extraction in the periphery (Bhambra 2007). These inscriptions have predicated a renewed emphasis on techno-modernist challenges of integration and connection to different trade and macroeconomic systems in neighboring countries. This is because competing colonial powers had little interest in building regionally connected economies (Chingono and Nakana 2009). Furthermore, the state of infrastructure that had been built was generally very poor and had in the main been deployed to serve colonial extractions in which "rail lines were linked to importing finished goods and extracting resources and agricultural produce and getting them to the ports for shipment to Europe" (Dienel 2016, 166). These legacies remain an ever-present dynamic in the economic performance of the region, with the German development agency, GIZ, estimating that "poor infrastructure cuts national economic growth by 2 percent every year and reduces productivity by as much as 40 percent" (GIZ, n.d.).

Focus on the active forces unleashed through colonial-capitalist histories on these regional infrastructure geographies is vital to any contemporary analysis of corridor initiatives. Understanding new corridor developments must therefore approach these technological formations through the "imperial durabilities of our times" (Stoler 2016). As Kimari and Ernstson (2020, 825) argued, analysis must connect to the "inherited material and discursive scaffoldings that remain from the colonial period." This means focusing on the sustaining of modernist planning discourses of integration into the world economy. It draws our analysis to how these have mutated into the contemporary era through the logics and imperatives through which various actors design and implement corridors.

However, to interpret this new era of proliferating development corridors only through the colonial logics that predicated investments in extractive railways, roads, and ports would be a mistake, "as a mere reiteration of imperial practice would mean to impose a Eurocentric interpretation on contemporary African history and subsequently disregard African agency" (Aalders 2021, 2). Rather, as set out in chapter 2, approaching the contemporary deployment of infrastructure corridors means to understand that these projects remain steeped in and mutate out of colonial histories and logics of trade routes, technologies, and economic relations with the world focused on various types of integration into a global system of racial capitalism but not entirely defined by such processes. As Cowen (2017 n.d.) argued, "Infrastructure is not only a vehicle of domination and violence. It is also a means of transformation." We might therefore (re)frame this new wave of investment into infrastructure corridors in recent years as both a rupture from the logics and rationalities of colonial/neocolonial extraction, and a continuation in new forms such as the competing infrastructural interests of China and the West encompassed in the "New Cold War" (Schindler, DiCarlo, and Paudel 2022).

CORRIDOR GEOGRAPHIES AND OPERATIONS

Infrastructure corridors have (re)emerged as new or repurposed networks of road and rail, associated pipelines, new cities, repurposed and expanded ports, and massive investments in digital and energy-generation infrastructure. There are at least thirty such cross-border initiatives in sub-Saharan Africa planned, under construction, or in operation, as well as ongoing attempts to integrate economies and territories through associated megasized infrastructure projects such as the Trans-African Highway network (Cupers and Meier 2020) and a myriad of national and sub-national scale initiatives.

I contend that there are two primary factors that are generating this proliferation of corridors. First, in the last decade, growing prosperity, state power, and increasing financial capacity have occurred in many African states. This has been predicated on the potential for national governments to enable expanding resource frontiers, new logistical geographies, and urbanization. Initiatives have built on strategic national and regional planning objectives that have led to new "infrastructure-led development" (Schindler and Kanai 2021). These new waves of projects are often centered around logics that find expression across urban space and the operational hinterlands of the urbanization process. Growing levels of finance, including new debt-based loan financing available to African nation-states, mean such plans are increasingly being materialized and delivered around the aspirations of the African Union's Agenda 2063.

Critical to these ambitions is the Programme for Infrastructure Development in Africa, which seeks to facilitate integrated infrastructure systems that spur growth and economic development. This will be activated through working toward the free movement of people, capital, goods, and services and bold infrastructural plans for the future, including the African High-Speed Train Network and Pan-African E-network, alongside vital growth in power generation by 50 percent. Cheikh Bedda, Director for Infrastructure and Energy at the African Union Commission, said, "Only by scaling up investments in corridor infrastructure could African countries participate in, and benefit from, today's integrated and digital global economy" (African Union 2017). And Edmond Were (2019, 566) has drawn attention how these projects represent a "postmodern Pan-Africanism" pointing to the "competitive race amongst the partner states and subscribe to the African Renaissance mantra but push countries into debt entrapment and protracted dependency."

Second, beyond the region, rising geopolitical ambitions and surplus finance-seeking accumulation opportunities establish a further financial basis for investment in large-scale infrastructure. Patterns of global trade and investment are undergoing historic restructuring, in which the power of North Atlantic–dominated capitalism wanes and declines and the Indian Ocean becomes a new locus of finance and power (Newhouse and Simone 2017). Such transformations are increasingly tied to Africa economies, with the Belt and Road Initiative (BRI) being the foremost economic force shaping infrastructural futures in the region. The BRI has generated or proposed massive infrastructural demarcations across land and ocean, with forty-two African countries having agreements or understandings with the Chinese government by 2020. It has opened up opportunities for African national governments to access new types and forms of infrastructural financing beyond the imperatives and conditions of Western governments (Oqubay and Lin 2019; Power, Mohan, and Tan-Mullins 2012), including at an urban or city scale (Goodfellow 2020).

The effects of these investments on urban life and the making of the Infrastructural South are starting to be explored through a growing set of studies on large-scale infrastructures (A. Demissie 2018; Enns 2018; Kimari and Ernstson 2020). As Simone (2010, 22) made clear, "Urbanization in certain sub-regions certainly is propelling new forms of regionalization and the gradual integration of national populations into regional domains, and is marking out widening corridors that expedite new economic synergies." With corridor plans incorporating more than 40,000 km of upgraded or new routes, it is evident that the techno-environments of the corridor are critical to the unfolding geographies of the third wave of urbanization.

In analysis that I undertook on the thirty corridor projects in the region, there was significant variegation in the objectives, purpose, financing, and technologies in operation. However, there were also some important commonalities that are worth drawing out. Intense and growing logistical hubs in port cities were nearly always present. Some corridors are directly connected to each other. For example, the West African Growth Ring proposals span three separate transportation and economic development corridors between Ouagadougou and the coastal cities of Abidjan, Accra, and Lomé, and remain aspirational and seeking finance. Operators have already begun activities on other corridors such as LAPSSET in Kenya, connecting to South Sudan and Ethiopia, or the Djibouti-Addis Ababa Corridor, connecting the Horn of Africa port city to Ethiopian hinterlands. The Walvis Bay corridors span a range of routes in southern Africa, incorporating and integrating Angola, Botswana, DRC, Namibia itself, South Africa, Zambia, and Zimbabwe. These routes include the Trans-Kalahari Corridor, the Trans-Caprivi Corridor, the Trans-Cunene Corridor, and the Trans-Oranje Corridor, all connecting at a regional gateway of the Port of Walvis Bay. Corridors sometimes overlap, with historical attempts at continental scale integration, such as the seven Trans-African highways, covering more than 55,000 km and first proposed in the 1960s to switch away from histories of colonial extraction and as "African leaders imagined infrastructure as a vehicle of Pan-African freedom, unity, and development" (Cupers and Meier 2020, 62).

The estimated cost of operationalizing these corridors amounts to more than USD 100 billion. A range of financial actors, from the World Bank to the African Development Bank through to National Governments within Africa and beyond, most prominently China, but also including actors such as the Japan International Cooperation Agency and the European Union, are involved. All are busy preparing sometimes rival spatial development plans and investing in new infrastructure to enable these corridors to become operational. Financial support, especially for feasibility and pilot studies, tends to emerge through African Development Bank–led programs such as the New Partnership for Africa's Development Infrastructure Project Preparation Facility (NEPAD-IPPF). However, the vast sums mobilized to operationalize these infrastructures depend on the purpose of the corridor. For instance, the USD 4.2 billion cost of the 1,070-km Chad–Cameroon Petroleum Development and Pipeline Project was primarily (95 percent) funded by a consortium of oil companies that would reap the profits from the extraction, whereas the USD 7.6 billion cost of the Central Corridor,

East Africa, is being funded through loans from the Chinese bank Exim to the Tanzanian government.

Anchored in Kenya, the LAPSSET corridor offers an interesting case of constellation of financial arrangements involved in shaping these infrastructures. LAPSSET is supposedly "the first largest Game Changer Infrastructure Project the government has initiated and prepared under Vision 2030 Strategy Framework, without external assistance" (Government of Kenya, n.d.). Through the Presidential Order Kenya Gazette Supplement No. 51, the government established the Corridor Development Authority in 2013. The African Development Bank Group estimated that its effect on "economic growth might even range between 8 percent to 10 percent of GDP when generated and attracted investments finally come on board" (ADBG, 2023). A myriad of financial mechanisms and actors invested up to USD 25 billion in the infrastructural corridor. This finance included a USUSD 1.5 billion contribution from the Development Bank of South Africa. However, the Kenyan state has been the primary investor, spending nearly 16 percent of its 2016–2017 national budget on the project. Additionally, China has provided the loan-based support for this national spending.

The entanglement of Chinese finance into Kenya's corridor ambitions is extensive, as with many African governments involved in these large infrastructure projects. It includes funding of the Standard-Gauge Railway from Mombasa to the border of Uganda through a USD 3.2 billion loan and concession for the China Road and Bridge Corporation, via its Kenyan subsidiary Africa Star Operations, to run services. By late 2020, the Transport Committee of the Kenyan Parliament was urging the government to renegotiate the terms of the loan, as the COVID-19-induced economic crisis left the country struggling with an increasing debt burden. It is not surprising that by 2020, China owned 21.3 percent of Kenya's external debt, including 72 percent of its bilateral debt (Africa Check 2018), emphasizing the importance of the BRI in enabling national governments in Africa to secure loan based finance for investment in infrastructure. Despite decades of austerity across public budgets imposed through the SAPs across more than forty countries in the region, it is notable that national governments can now invest again in infrastructure through finance from beyond the Western-dominated institutions of the IMF and World Bank. And these flows of finance from and to China extend beyond the actual financing.

For instance, the port expansion at Lamu incorporated thirty-two berths—a vital prerequisite for the LAPSSET project. The China Communications Construction Company has awarded the USD 689 million contract by the Kenyan government to implement this works.

Corridor projects such as LAPSSET and others clearly point toward a new era of large-scale infrastructure projects in Africa after the malaise and underinvestment of the structural adjustment years. This surge of plans, proposals, and actually constructed projects echoes both the postindependence modernizations and colonial extraction eras in the ambitious regional-urban planning of new logistical geographies. It proceeds in a very different context, particularly at the urban scale.

NEW URBAN REGIONAL GEOGRAPHIES OF THE CENTRAL CORRIDOR

The Central Corridor is part of a strategic planning and investment effort to transform Tanzania into a crucial hub in connecting the East African hinterlands to the Indian Ocean and the promises bound up in the Maritime Silk Road and the wider USD 1 trillion BRI (Enns and Bersaglio 2020). With various rivals for facilitating these logistical flows, primarily the Kenyan LAPSSET project, the Tanzanian government needed "to strategically position the Central Corridor as the most efficient in East and Central Africa so as to contribute positively to poverty-alleviation programmes in member states" (Central Corridor Transit Transport Facilitation Agency 2018). Tanzanian ambitions to capture a significant share of the 90 percent of Ugandan ocean-going trade that previously flowed through the Northern Corridor and on to the Kenyan port of Mombasa hinged on the promise to finance necessary infrastructure and to speed up the flow of goods to ocean-bound traffic from Kampala.

With multiple developmental pressures, Uganda has, like many other African countries over recent years, sought to use infrastructure to enable the technological and territorial reordering of its economic spaces toward enhanced growth (National Planning Authority 2013). This has signified a shift by the government away from investments in basic service provisions toward new, regional-economic infrastructures. In late 2015, I met with a senior advisor to the president's office in the gardens of the Mount

Elgon Hotel in Mbale. He told me of his recent trip to China to secure a loan. The advisor told us about the president's rationale for deciding priorities for national development and growing the Ugandan economy. The focus was to be not "on the democratization of poverty but on big infrastructure such as roads and energy."

The making of the Central Corridor was driven by the Tanzanian government but is also integral to the wider economic ambitions contained within Uganda's Vision 2040 Plan. The plan imagines an infrastructural advanced country made up of "a high-tech ICT city, 10 new cities, four international airports, national high-speed rail and a multi-lane road network," all to propel industrialization and the growth of a service-based economy. As the plan states, "Ugandans desire to have world-class infrastructure and services, and modern technology to improve productivity and production" (National Planning Authority 2013, 10).

Making such national economic development possible involved a series of actions, many beyond Ugandan borders. This included establishing or integrating a range of technologies, sites, and systems into the Central Corridor's larger network and the myriad of new projects Tanzania has been pushing. These included large investments such as improved, expanded facilities at the Port of Dar with globally standardized technologies and the construction of a standard-gauge railway connecting Dar with Port Mwanza on Lake Victoria. It included upgraded ships, the MV Kaawa and MV Umoja, improving speed to seventeen hours per crossing and with a capacity of 880 tonnes per vessel as well as improved safety (brought into sharp focus after the 2005 sinking of the MV Kabalega).

In Uganda, the government has spent USD 10 million on the rehabilitated Port Bell, the repurposing of the track 12 km to Kampala that would open up hinterlands beyond, and locomotives to ply the route. Interestingly, some of this finance was also secured from the European Union. All these components of the multimodal transportation system have been necessary to fulfil the economic promises of the Central Corridor[1] (Central Corridor Transit Transport Facilitation Agency 2018). Without the railway track through Namuwongo, the planning and finance required to make viable this corridor would have remained unfulfilled. It meant that the urban space used by residents in Namuwongo for everyday activities, including the infrastructures of social reproduction such as shelter, enterprise, or visiting the toilet, would now be allocated to the Central Corridor.

The future of the Central Corridor is uncertain in that its logistical and financial promises are far from guaranteed. While its ambitions might involve the shifting of various circulations of Ugandan goods and trade away from the previously dominant Port of Mombasa in neighboring Kenya and the acceleration of these movements out of Uganda, this does not mean that the deployment of this multimodal system fulfils the promises and hopes of planners, politicians, and economists at various government ministries or whether it will generate sufficient income to repay loans to the Exim Bank. The investment in this corridor is a calculated speculation to improve the economy of Uganda (and Tanzania). It is a form of speculative, technological deployment, which seeks to draw in various economies, circulations of resources, extractions, and goods to achieve the economic objectives established in the Vision 2040 plan. Repurposed spaces such as Port Bell and the reactivated railway offer a representational power to convey to the world that Uganda has an open economy in order to secure foreign direct investment through conveying the message that the country is open for business and can fulfil tight logistical deadlines in delivery of goods and services.

Reinvigorated geopolitical relations established to create a cross-border 1,400-km pipeline for oil production in Hoima, western Uganda, are also crucial to understanding the deployment of the Central Corridor. Accessing global markets from the Port of Tanga, plans project the East African Crude Oil Pipeline to carry 2.2 billion barrels of recoverable oil reserves in western Uganda to global markets at a cost USD 4.4 billion to construct. The China National Offshore Oil Corporation is the primary beneficiary. Deployment of the pipeline-facilitated extractive flows established the political and investment conditions for further technological and territorial alignment. Techno-environmental integration is produced through the twin infrastructures of oil and transportation, enabling circulation between East African neighbors. Governments weave infrastructural layers across material, political, territorial, and economic conditions that extend to encompass new configurations.

The Central Corridor remains a speculative project predicated on the promise of infrastructure because it competes with more established regional infrastructure plans, including the USD 25 billion Kenyan East Africa Railway Master Plan (Sambu 2008) or the LAPSSET Kenyan mega-infrastructure project (Bremner 2013), both of which seek to establish the Kenyan coast as the crucial staging post for the circulation of Ugandan goods, extractions, and people into the world economy. Other uncertainties infuse the potential futures of this infrastructure. These include economic instability across the region or diplomatic tensions between Tanzania and Uganda. Despite the promises and hopes of national economic planners that are materialized through various plans, investments, and strategic calculations involved in making the Central Corridor operational, the outcomes of transnational, transoceanic connection and the attendant economic growth are never certain. Incalculable uncertainties range from regional economic instability to more competitive options being opened to technological failure or interruption to political contestation.

The presidential advisor I met confirmed these new infrastructure-led forms of diplomacy, investment, and shift eastward in search of finance. He described how the delegation met the Chinese government, who were very welcoming and obliging, asking what Uganda would want as a sign of friendship. Eventually, the governments made an agreement for a loan deal for USD 1.9 billion from the Exim Bank. This was intended to construct Karuma and Isimba hydropower projects, adding nearly 800 MW to the country's electricity capacity through a twenty-year repayment plan. The advisor was surprised to find that the Chinese delegation was also interested in what other infrastructure investments Uganda would want, such as roads. In exchange for and beyond the loan conditions themselves, the Chinese wanted access to the new oil fields in the west of Uganda, and this was part of the final partnership deal, reflecting a similar story across Africa in recent years. Chinese attempts to control extractive infrastructures focused on natural resources, partly through providing loans for other forms of infrastructure, may become the dominant financing model.

COLONIAL INSCRIPTIONS

Before considering the urban dimensions of the Central Corridor in Kampala, it is important "to track how colonial processes continue to scaffold" (Kimari and Ernstson 2020, 3) in this project and how they shape the

6.2   The Central Corridor promises easy access for Uganda to the Indian Ocean.

techno-environment made up around it. The construction of Port Bell and the train line to Kampala were first developed by British colonial authorities as a later spur for the Lunatic Express railway line, a near 1,000-km endeavor, involving 30,000 laborers to deploy a strategic rail link between the Port of Mombasa and the East African hinterlands with the aim of opening up new cotton growing territories. This colonial transportation corridor required legislation in the form of the Uganda Railways Act (1896). The colonial authorities built it, despite "Conservative and Liberal governments of the early 20th century" being "wary of spending large sums in the Empire, particularly in Africa" (Robins 2016, 82). Indeed, the British politician Henry Labouchère described it as a "gigantic folly" (Hansard 1900), demonstrating its speculative nature and uncertainty over its capacity to create value for the colonial power.

The specific demands of the emergent cotton production in Uganda were integral to the establishment of Port Bell and the railway line from Kenya. This was driven less by the UK government and more by the British Cotton Growers Association in search of ever-increasing accumulation opportunities, as well as the support of Winston Churchill, who visited in 1907. O'Connor (1965, 47) argued that the spur was driven "chiefly through the establishment of cotton, this part of Uganda was thereby given a lead in economic development which it has maintained ever since. The growth of Kampala as a commercial center was assisted by the building of a six-mile railway to Port Bell in 1913, which removed a heavy burden of porterage and relieved congestion which was holding up import traffic."

The promise of infrastructure to aid economic growth (as a colonial node in the British Empire) is also clear in the words of the general manager of the line in 1927, who argued, "I believe that the line will attract fresh capital to Uganda; will produce additional traffic: . . . and will stimulate general development in the whole of Uganda itself" (in O'Connor 1965, 56). The railway line itself started life as a driven Ewing system, based on a monorail line, before conversion in 1913 to a meter-gauge railway that has stood ever since. By 1930, the port, busy with traffic from across Lak Victoria was also a landing point of the air route run by Imperial Airways, becoming a truly multimodal transportation node in the geographies of British colonial extraction and rule on the continent. Recent investments echo the inscriptions of previous eras of infrastructural governance and the promises of development for Uganda, but at what cost to those living in its way?

THE INFRASTRUCTURAL LIVES OF THE CENTRAL CORRIDOR

To understand better how these massive investments are transforming urban life, we return to the popular neighborhood of Namuwongo. The national elections that brought President Museveni back to power in 2016 had paused efforts by the Uganda Railways Corporation to clear a 30-m zone either side of the railway track that ran down the settlement. This was required by various operational standards and procedures relating to the reactivation of the train line to Port Bell, and the broader implementation of the Central Corridor. Deemed as politically damaging by the ruling National Resistance Movement (NRM) party in the lead up to the election, the community enjoyed a moment of respite from the constant worry and uncertainty about whether they would lose homes and businesses that lay inside the demarcated railway line. As "Moses," a local resident, pointed out during campaigning, "The president said Namuwongo will not be demolished during the rallies, so I think we are still safe." But this pause was not to last. In the days after the 2016 election, a local leader of the main opposition party at the time, the Forum for Democratic Change (FDC), was attendant to the increased danger that residents faced. He explained, "The future of Namuwongo over the next six months is uncertain, and people's lives are at stake. Since the run up to the election last year, the demolition has been at a standstill, but it's clear people will suffer even more in the future."

As Joel and I proceeded down the train track, we saw how everyday life continued with little outward sign of concern about how a rehabilitated train line, connecting Kampala to the Indian Ocean, might place years of work, savings, calculations, and collaborations of residents at risk of being displaced. Local people had established businesses all along the ruins of the colonial-era standard-gauge track. An assortment of retail stalls snaked past the market and into the nearby industrial zone: replica football kits of various European teams and the improving Ugandan Cranes at one stall, supplies of charcoal that had arrived from the nearby islands of Lake Victoria at another, various phone charging points, several "VIP" toilets set up by entrepreneurs for passing trade, and the ubiquitous "Rolex" stands selling Uganda's favorite snack.

A popular economy was in operation—a social infrastructure that residents depended on to navigate life in a resource-poor context. That is not to say that residents did not have long-held fears, having already experienced an attempted and partial demolition a couple of years earlier, especially as we continued with our research as the elections finished and business in the city was getting back to normal. We met "Joy", who was busy after the election passing through various gatherings to find out the day's news. She was quick to recognize the unfolding danger of making the railway functional again, telling me, "The railway is taking away the land. So, many of us will be evicted. Right now, we have no future here because we know any time, we shall be told to leave."

It is nearly always the global-facing infrastructures of connection, extraction, and circulation that are prioritized in these conflicts over urban space usage. The demands on urban space for national economic development take precedent over those of the urban poor. "Patrick," another resident in Namuwongo, was clear about such governing of urban space: "There

6.3 The railway line through Namuwongo.

is no future in Namuwongo if this government is still in power. This is a government that doesn't mind about the poor people. They just think of satisfying themselves and their own people, while the rest of you should just find your own way out."

SHATTERED SOCIAL INFRASTRUCTURES

It is worth setting out the profound impact on fragile social infrastructures and the incremental techno-environments that operated along the train track in more detail in relation to the deployment of the Central Corridor. A range of social infrastructure to navigate the everyday challenges of life in the city have been developed by households in Namuwongo. This is resonant for those whose homes or businesses are on Uganda Railways Corporation (previously Rift Valley Railway) land or the wetlands managed by the National Environmental Management Agency (NEMA), which authorities deem as ineligible for investment of basic infrastructure. For instance, KCCA cannot establish state-run systems for sanitation provisions in the neighborhood in these spaces. In response, a "grid" of social infrastructures that are incremental, heterogeneous, and peopled are operated by residents to enable social reproduction.

Social infrastructures operating across Namuwongo, as elsewhere in urban Africa, are often simultaneously place specific and extended. They are shaped by and rely upon localized techno-environmental conditions and social relations, alongside connections to people and places further away. Living next to neighbors builds up trust and understanding that allows for favors, support, and help that are less likely to be given to a stranger. These types of support take time to build up and sustain and cannot be easily replicated in a new place or at a different moment. The spatial immediacy of living as neighbors allows for collective action to deal with everyday challenges together (although at other times may predicate conflict). Elaborate arrangements between various households often reflect particular techno-environmental conditions and the disposition, networks, and capacities of those involved, from building a toilet in the bit of land separating two households to avoid dangerous nighttime trips to a nearby facility, through to looking after children when illness occurs, to shared tasks such as disposing of garbage, building a walkway during the rainy season or extending a new electricity connection.

These necessary tasks are underpinned by a social infrastructure that are often most intense in urban spaces in which state-operated infrastructures services are most minimal. Social infrastructures span beyond the immediacy of neighbors and can encompass various networks across neighborhoods, cities, and villages. In a neighborhood such as Namuwongo, these social infrastructures are thick and yet fragile to impositions from outside such as logistical investments. To think in more detail about this tension between development and displacement that exists across new deployments of corridor infrastructures, I focus on one particular person who embodied many of these experiences.

Jennifer arrived in Namuwongo in 2002 after losing her job and with a family to support. Initially renting a room before using her various skills, entrepreneurialism, and capacity to create networks of support and economic connection, she built her own house in the neighborhood. Using two years of savings from her catering business, she set out and bought a plot of land along the railway. Even at the onset of this self-built housing project, Jennifer had to navigate the uncertainty of the land titles she had supposedly gained, with others also claiming they had purchased it. Her response after seeking advice was to begin construction immediately on a three-roomed home as the best strategy through which to sustain her claim. From paying for this land in early June, they put up the basic structure by mid-August. The claim on this land meant a future for her family, a space to live in the city, and, more immediately, a physical structure to hinder others from seeking the land.

The build was initially basic; windows, floors, and other essential parts of the house remained incomplete, and the family had to negotiate mosquitos and snakes during those first months. Slowly, builders assembled the various components. Windows were purchased from a widow down by the wetlands. Within the year, Jennifer had plastered the outside to protect against the torrential rains of the wet season. After another year, Jennifer invested in a cement floor, which made the home feel more permanent. The piecemeal or incremental upgrading of this structure into a home is a common experience for millions of urban dwellers. It provided for the housing needs of Jennifer's family, a space to live in the city, and a place of convergence for the various social infrastructures that this group of residents were establishing.

Jennifer enjoyed the new house. It meant she didn't have to pay rent, and it felt like a big step forward for the family—a base in which they could carry on progressing. The building provided a physical space in which Jennifer could coordinate an infrastructure of social reproduction that she put to work for the benefit of her family, friends, and wider community. This social infrastructure extended far beyond her family. She was a chair of a women's group based in the area. It was a group that made beads, groundnut paste, and craft shoes. Jennifer started the group, which grew to forty-two women. There was also a savings element, with the women asked to save UGX 1,000 (¢30) each week to enable plans to be activated. Jennifer was also an active member of the Namuwongo Christian Center, a Pentecostal congregation whose building was also along the train track, which provided various community services.

Things changed in July 2014 when the municipality demolished structures along the track to begin the process of bringing back railway services and to prepare for the operationalization of the Central Corridor. Jennifer's house was close to the train track, perhaps less than 10 m up the slope and clearly in the zone claimed by the Uganda Railways Corporation. The structure's closeness to the line had been useful for Jennifer in securing passing trade on various business ventures she was involved in, but it now meant that its very future was at risk of being destroyed. On this occasion, the authorities did not demolish her home. The Nakawa High Court temporarily halted the eviction of residents throughout Kampala and living close to railways deemed as important strategic infrastructural assets for national development. This injunction was too late for many families in Namuwongo, whose homes and businesses they had demolished without warning. The demolition occurred in the middle of the night, without warning and leaving little time to rescue sentimental or valuable artifacts. The demolition in 2014 was traumatic, impacting the uncompensated evictees as well as their friends and neighbors. There was little sympathy from the authorities. "Mary", a planner at KCCA justified the lack of compensation, saying, "We are not going to pay people with taxpayers' money to move off the railway line and to people who don't contribute to the cake. So, we say, you must leave the railway line without payment. That area is gazetted and has been for a long time."

"Mary" also highlighted how the Ugandan state was unconcerned about the loss of housing for Jennifer's neighbors and the displacement being generated through the deployment and operation of the Central Corridor. She explained the demolitions were "complex, [and] it's no longer our mandate to provide shelter to citizens, it's now up to citizens themselves." The municipality failing to offer compensation or to regard residents' lives and their claims to the city in the planning process is an experience shared by urban dwellers from Karachi to Kinshasa.

This experience shook Jennifer. It was not long before the family moved out of this self-constructed home; she decided they should not live under the specter of a nighttime raid by the authorities that would put them in physical danger and leave them anxious and feeling insecure. Another resident highlighted this anxiety of living under these precarious conditions, explaining in 2016, "I can't tell when exactly demolition will take place, but I have that strong feeling that Namuwongo will be demolished." Jennifer calculated that the future risk of demolition was too high a price to stay around. She did not want to be exposed to the violence of the state and forced displacement. If the bulldozers and demolition crews had not made it to her house this time, then there was no guarantee that she would be so lucky again. They found a house further down the railway line, toward Port Bell and well outside the Uganda Railways Corporation delineated zone. She would rent her old home for as long as it was standing and use this revenue to help pay toward the rental costs of their new home. But this also brought problems, from the extra money required to pay the landlord through to the dislocation felt by losing neighbors and friends that had been part of everyday life for Jennifer over the previous five years.

Despite fears by many in the local community of state-driven clearance, contestation over the attempts to make the railway line operational was noticeable. This did not take the form of direct resistance to the construction and rehabilitation of the railway. In a country ruled by an authoritarian government and with a police force and military used by a president unafraid to use brutal force to suppress dissent, physical resistance to demolition and displacement was unlikely. Instead, in the lead up to the national elections, residents in Namuwongo, like urban dwellers across Uganda, supported opposition parties such as the FDC, hoping that after thirty years of power by President Museveni and the NRM, a change in government might predicate a more democratic and fairer Uganda. Their hopes were to be dashed again. After the 2016 election, the Uganda Railways Corporation set out to define the zone through physical markers, placing posts with the letters "UR" painted down them and on either side of the track. "Anna," a resident, let me and Joel know that "the demolition plan is close because I see the railway mark planted along the railway. One of them is just next to my door. Most people in Namuwongo are just on standby, waiting for the day when they will say go and they go."

"David," another resident, told me and Joel that the informal arrangements that seemed to govern the use of this railway land were now broken and that continuing to inhabit such a space was to experience daily uncertainty about the possibility of staying put. He said, "Right now, we have no future here because we know any time we shall be told to leave. A person like me, I have no plan right now, and I never thought something like this would happen so soon. When we were building here, we asked the railway officials, and they gave us the go-ahead to build here. So, now, if they are evicting us, then we have no choice."

6.4 Uganda Railways Corporation posts in Namuwongo.

And then, for a while, nothing happened. The posts remained along the track, and the demolition failed to materialize for some weeks and months of uncertainty until one day it happened. The demolition team set to work to make the track operational, demolishing any structure that stood in the way. The train services that connect Port Bell and central Kampala after a pause of decades resumed, celebrated by the government and other partners as another step forward in achieving the Vision 2040 plan. Residents had lost their homes and businesses. They had been dislocated from the social infrastructure that had sustained life in the city. The relations and makeshift arrangements were destroyed.

In postcolonial fiction such as Thiong'o's (1977) *Petals of Blood*, there is a rich literary tradition that portrays modernity for Africans as the coming of a road or railway track and the ways these investements shattered social worlds. The Indian Ocean–facing Central Corridor and the social infrastructures that underpinned everyday life in Namuwongo were two types of networks that required usage of the same urban space and very much reflect Thiong'o's storytelling. Furtheremore, a key similarity exists in that both of these infrastructures are in effect speculations that attempt to control the circulation of various socio-natures into a way to generate income, to sustain various scales of economic activity and to make life viable in the city. Writing about what they term "lived infrastructure," Graham and McFarlane (2014, 19) argued, "people engage in speculative transactions through which they attempt to anticipate the actions of others." We can extend this to various scales across and beyond the city that show the uncertain futures that suffuse corridor techno-environments.

CORRIDOR POLITICS AND IMPERATIVES

What are the politics and imperatives that surround new corridors, and how do they connect to the making of the Infrastructural South? Given the intensification of proposed corridors into national and regional development visions across the region, such investments seem likely to kick-start a new era of economic growth across urban Africa. These initiatives show the historical and contemporary power of the corridor as a conduit of force to make and remake multi-scalar geographies that reflect the place and position of African states in the world economy.

Three sets of political imperatives emerge in relation to the techno-environments of the corridor. First, the emerging prominence of China in the financing of new corridors has allowed newly assertive national governments to finance long-held infrastructural ambitions, but it also opens up a series of uncertainties about the capacity of these states to repay loans and the consequences of failure of these schemes. The risk of the Tanzanian government defaulting on its loan from the Exim Bank of China for the Central Corridor has led to fears by some that defaulting could hand control over this vital national infrastructure to China, mirroring the experience of the Port of Hambantota, Sri Lanka, being handed over to China control in late 2017 after the government failed to repay a USD 1.3 billion loan. And in Uganda, 75 percent of bilateral debt owed is now attributed to the Exim Bank (Pasaogullari 2019). Much of the USD 95.5 billion that China has lent in Africa between 2000 and 2015 primarily focused on infrastructure, and the East Africa region alone now owes China USD 29.4 billion in infrastructure loans (Nation 2020). Fears of a new iteration of the "debt trap" (Payer 1975) are present, even as they are yet to be, and indeed may never be, realized.

Perhaps of more importance in thinking about how the Infrastructural South is being (re)made are the decisive shifts of extractive and mobility infrastructure of the region away from the old colonial powers. They now extend toward the emergent economies of the Indian Ocean. The rise of China has emerged as the capacity of the West to intervene and shape infrastructural geographies and operations has begun to fade. Africa's urbanization is now being engineered into a new global geography of trade and logistics.

Second, focus is required on the inscriptions that remain bound up with the geographies and operations of the techno-environments of the corridor. In thinking through these more than "imperial durabilities'" (Stoler 2016), we are left questioning how inscriptions from previous eras of infrastructure investment, colonial and otherwise, guide an extractive logic that resonates into the contemporary era. As I showed through a focus on the Central Corridor, much of the infrastructure required for operation in Uganda had been established in the colonial era. This was a history of British investment focused on speculative projects for the Empire to accumulate from the natural resource potential of the region. Postcolonial rehabilitation from the Belgium authorities to enable continued extraction from the DRC after colonial rule followed. These inscriptions have now been updated and standardized for a different era of global trade and logistical circulation. However, new itinerations of the corridor remain tied to the technological and territorial logics of the past.

Chinese finance now primarily facilitates contemporary financing of infrastructure corridors. However, these flows arguably remain within similar logics of extraction and logistical flows that animated the intentions of the Western powers. These geographies are "scaffolded" (Kimari and Ernstson 2020) by colonial histories and relations with the West and the role of African cities within these unequal imperial networks of global trade. It generates important questions about how much has changed and transformed beyond the technologies, finances, and geopolitical relations now at play and therefore what capacity long-independent states such as Uganda or Kenya now have to shift these geographies and operations of the corridor outside of these logics and circulations. Like in the modernization era of the postindependence era, African states are once again at a crossroads to move beyond the political economy of extraction, but the techno-environments of these extended infrastructure systems such as the Central Corridor or the Northern Corridor through Kenya suggest business as usual.

Any kind of shift would require a definitive rupture from the past. As Achille Mbembe (2018, 4) argued, "we have to bring down colonial boundaries in our continent and turn Africa into a vast space of circulation for itself, for its descendants and for everyone who wants to tie his or her fate to our continent." The new wave of corridors and the techno-environments they are engendering across and beyond urban Africa reinforce the operations and logics that shaped colonialism in the region into the contemporary era and suggestive of the need for Pan-African responses that think anew about these movements and flows outside these rigid geometries.

Third, and particularly important at the urban scale, are the displacements and dispossessions wrapped up in the deployment of corridors, stemming from the extensive demands for urban space of these speculative, logistical technologies. Displacement is not just experienced in Namuwongo. For instance, in Kenya, both the popular neighborhood of Kibera, Nairobi, and communities in Lamu have faced a similar fate as new road infrastructure and port expansion have proceeded. Similar processes of corridor-led dispossession are occurring across the ever-shifting urban worlds of the region. To operationalize a corridor requires restructuring of existing urban spaces and functions, and often this is at the cost of those already living marginalized lives.

The prioritization of corridors in national plans may intensify unequal access to infrastructure already critical for shaping urban life. It produces conflict between the deployment of the requisite systems to enable human and nonhuman circulations and flows across towns, cities, and their hinterlands and its displacement of existing grids of social infrastructure. As Enns (2018, 106) argued, "corridor development enables certain flows of capital, commodities and people to move easier across space, while introducing new forms of spatial exclusion and immobility for others." These are the dialectical effects of infrastructure between attempts to enable new forms of connection and the potential ways urban populations become disconnected through such initiatives. As Rao (2014, 39) notes, "To talk about infrastructure is to invoke both the promise of a future as well as imminent trauma." The promise of new, accelerated logistical connections to the global economy for the African region often takes precedent,

6.5   The railway line through Kibera.

as they have done before over the everyday spaces that urban populations depend upon. Its imminent trauma is the displacement and dispossession required across delineated zones incorporating a whole ensemble of social infrastructure that helped residents navigate everyday life. Apostolopoulou (2021, 831) termed such transformations as "infrastructure-led, authoritarian neoliberal urbanism" that "engender a new stage of revanchist and authoritarian urban development that deepens spatial fragmentation, territorial tigmatization and social segregation." The techno-environments of the corridor empower a new phase of urban uncertainty, especially as many of them may remain mere plans or are only partially realized. For instance, the Ugandan government abandoned attempts to further expand Port Bell for a USD 200 million project to create a new lakeside port in Bukasa. This produces lived experiences of uncertainty for those least equipped to navigate such turbulence generated from attempts to fulfil the promises of national economic growth and development.

7

INFRASTRUCTURAL CATCH-UP "OFF THE MAP"

STRIKE/PROTEST

Sydney and I arrived at the sizeable landfill site early in the bright morning with a couple of municipal officials who had agreed to show us around. As "Tony", the facility manager, led us from the gates toward the new building, he spoke about how this restructured operation promised the eastern Uganda town of Mbale more efficient management of waste. This was a shortcut to achieve the infrastructural catch-up. After years of breakdown and dysfunction, ad hoc collection processing of the garbage had been left to dozens of waste pickers, who congregated, sifted, and sorted each day. We saw a couple of workers toiling their way through waste, even as others were not doing much. They were separating organic matter before it was bound up in piles to be transformed into compost destined for local farms. This investment looked like a winner for the town, with new collection points, a new vehicle, and an upgraded waste facility. It promised to address waste-processing issues to improve the local environment, health, methane emissions previously emitted from the garbage, and local employment through new positions for the collection and sorting teams. The tour and demonstration concluded. We said our thanks and headed out through the industrial zone and back into town.

The next morning, we showed up at the site again, this time unannounced. A very different scene was at play. The previous day had been a performance for my benefit, an out-of-town *Mzungu*. After speaking to some workers without the municipal officials present, they informed us they were on a wildcat strike over unpaid wages. They assembled every day on site, hoping that the private company, given the contract by the municipality and who one older worker described as "someone behind us, taking money like a ghost," might pay up. By the time of our visit, things had become really tough for the workers. Unpaid rent and the ability to purchase food and other essentials was becoming a daily struggle as credit lines dried up at nearby stores and landlords became increasingly inpatient. Other tactics were now at play for the workers to survive, including collecting plastic bottles on site that might bring UGX 500 (¢10) or so for each sack.

Later, Sydney drove me round the town and the mounds of trash growing alongside the new waste-collection points, some of which were now named by frustrated, adjacent communities after local politicians. We found "Joseph," the councillor of Doko Cell (who had not had a mound named after him), surprised to hear the municipality claiming the system was working. He laughed and drolly asked, "You are informing me it is still operational?" "Joseph" talked about the difficulties facing some of his community who were taking part in this informal strike and how implementation of the restructuring had not gone as promised. Mbale still faced uncertainties over its waste-management, despite this new investment and its promises, and now the workers were revolting.

In September 2014, Julius Malema, former ANC Youth League firebrand and leader of the Economic Freedom Fighters (EFF) party, was due in the magistrates' court in Polokwane to face charges of fraud, corruption, money laundering, and racketeering. The night before, Melu and I, undertaking research on the city's energy networks, decided to attend a vigil being held by the EFF not too far away from where we were staying. It was to prepare for a large mobilization the next morning to coincide with court proceedings. We arrived at the Nirvana Civic Center, across Thabo Mbeki Street from the city center, at around 8:00pm, and we soon realized that this would be a long night for those assembled. Groups disembarked from buses, some from across Limpopo province, some from further afield, nearly all with their striking red EFF berets. We hung around for a few hours, talking to various

7.1   Piled-up garbage by a dumpster in Mbale.

7.2   Barbed wire outside the Polokwane magistrates' court.

people, many of whom made clear in no uncertain terms their dissatisfaction with the government's record of service delivery. Another procession of supporters slowly made its way into the hall. We realized it could be morning before Malema himself showed up, and we headed toward a nearby petrol station to pick up a taxi and return to our hotel.

The next day, as we made our way to the municipal offices for meetings, the streets of this small South African city were thronged with the red of the EFF supporters. Police vehicles barricaded direct access to the court with barbed wire across the junction. Malema addressed the assembled EFF activists, leading the song "Kgante motswalle waka, ke bolawa ke wena!"—a song about betrayal of an ally—before heading into the building as supporters carried on their determined singing. However, this assembly was to be short-lived, with the court immediately postponing the case until next year. The streets quickly emptied as supporters from near and far started their journeys home with no actual court action really taking place.

The EFF has risen to become a disruptive force in South African political life and especially through expressions of fury and anger against the ruling ANC. The night we joined the vigil, we could see the emerging, populist policies around the expropriation of natural resource monopolies had caught the attention of some South Africans. In Polokwane itself, many residents who were now EFF supporters had rallied against the ANC because of the costs of basic services, particularly electricity and water. The threatened imposition of smart meters in the city was a hot topic as this fiscally struggling municipality explored new technological solutions to increase its revenues and boost its budget. The meters had become a simmering point of anger for communities as implementation proceeded, with moments of protest and tension over the proceeding years. In 2018, residents in the township of Seshego had marched to the civic center, calling for the meters to be removed from homes as households struggled with sustaining flows of electricity. Polokwane's communities, like other urban areas in South Africa, were in a semi-permanent state of protest and mobilization against the state and its failure to deliver affordable infrastructure services. The brief but dramatic appearance of the upper echelons of the EFF in the city and the march to the center of the city by some of the Seshego community represented two different but connected politics of anger that now typified the urban infrastructural experience in South Africa (J. Brown 2015).

Mbale or Polokwane are perhaps not the first places that spring to mind when thinking about Africa's urbanization process and its infrastructural underpinnings. But like other towns and small cities, these are critical urban spaces of the Infrastructural South. The United Nations defines towns and small cities as urban agglomerations with populations ranging from 20,000 to 500,000 residents. As David Satterthwaite (2016, 5) established, "by 2015, around 196 million people lived in small urban centers, in sub-Saharan Africa—equivalent to almost half of the urban population and a fifth of the total population."

In this chapter, I revisit Jennifer Robinson's (2002, 531) notion of "ordinary cities" as a response to the "large number of cities around the world which do not register on intellectual maps that chart the rise and fall of global and world cities." Robinson highlighted the spaces included and excluded from urban theory making and argued for a reformulation of how such theorization proceeds. Recent years have witnessed proliferating work across urban studies that have shifted the production of theory from the paradigmatic northern cities, including work focused on infrastructure. However, these new ideas have generated knowledge claims of the South through a particular set of large or mega-sized Southern cities from Sao Paulo to Jakarta, Delhi and Mumbai to Beijing and, in Africa, Johannesburg and Lagos. This new set of "paradigmatic" metropolitan regions have become necessary sites of theory production in the challenging, problematization and move beyond Western-imposed narratives of urban modernity.

If these reorientations in global urban studies, mirrored to some extent in work on infrastructure, are to be welcomed, they arguably do not go far enough. A resident or visitor to a town or small city such as Mbale or Polokwane might reflect upon the limitations of thinking about urbanization in these spaces through experiences of large-scale urban agglomerations. Responding to this knowledge gap means to engage the Infrastructural South through towns and small cities that remain "off the map" of urban theory and policy (see Kanai, Grant, and Jianu 2018). Such "ordinary cities" (and towns) constitute a highly differentiated and distinct secondary techno-environment to that of bigger agglomerations (De Boeck, Cassiman, and Van Wolputte 2010; J. Robinson 2002; Rondinelli 1983) suffused with a particular set of infrastructural geographies.

LIFE BEYOND THE BIG CITY

Small and medium-sized urban areas, referred to as towns and small cities, form a significant feature of Africa's urbanization and have historically housed a greater number of urban dwellers than that of larger urban areas (Satterthwaite 2016). If total urban populations are more likely to live in bigger urban areas in the future, towns and small cities will remain integral to the urban experience as tens of millions of urban dwellers make these spaces their homes.

Historically, the colonial system and its logistical networks of extraction relied as much on the small outposts of administration, trade, and transportation as the larger centers of imperial power (Myers 2011; Owusu 2018). Sometimes, these settlements built on, erased, or expanded precolonial urban spaces; in other instances, they established new spatialities. In the postcolonial era, towns and small cities have expanded, fulfilling roles as regional centers for health, trade and exchange, flows of people, resources, investments, and, in some areas, aid to rural hinterlands. The particular urbanization imperatives being generated by these towns and small cities do not sit at the core of urban policy formulation, including in the planning of infrastructure services. As a UNICEF/UN-Habitat report (2020, 3) explained, "many of the urban development and governance interventions have focused more on primary and mega cities, presumably with the expectations of trickle down of social, economic and physical developments to other tiers of towns and cities."

In previous work with Cheryl McEwan, Laura Petrella, and Hamidou Baguian, we developed analysis of the growing intersections of climate change and urbanization in two "ordinary" smaller cities in West Africa: Bobo-Dioulasso, Burkina Faso, and Saint-Louis, Senegal. We found that much of the emphasis on responses to climate change at an urban scale was focusing on the large metropolitan regions, preventing an understanding of the diversity of processes, trajectories, and futures in smaller urban centers. We argued for a need to generate "context-specific knowledge, which takes account of geography, different climate change challenges, urban governance, and economic and cultural issues that in turn shape the economic development vulnerabilities and responses across 'ordinary cities'" (Silver et al. 2013, 675). We need to revisit knowledge production across the social sciences and urban policy, which prioritizes the large urban area. To do so means to take seriously the particular and generalizable characteristics of towns and small cities in how we come to think about the Infrastructural South.

There have been long, important traditions of thinking about these "secondary" urban spaces (Hardoy and Satterthwaite 2019; Klaufus 2010; Otiso 2005), even if there has been less attention on what this means from an infrastructural perspective. There are data pertaining to various sectors of urban service provision indicating a lag or deficit compared to larger cities in the capacity to provide safe, functional infrastructure. A study in South Africa considered a variety of indicators, including sanitation, water, and electricity provision, and concluded that "these cities had been in a less favourable position than metropolitan municipalities" (Marais and Cloete 2017, 187). This small city and town context very much chimes with the "infrastructural catch-up" developed in chapter 3 to describe the attempts to provide urban service provisions on the peripheries both inside and particularly outside the boundaries of metropolitan regions and perhaps under even more challenging circumstances.

The restrictive fiscal context faced by towns and small cities in relation to infrastructural investment, management, operation, and subsequent experiences by urban populations (Cirolia and Mizes 2019) is a defining underlying dynamic that establishes what I term a "secondary techno-environment." Reliance on grant transfers from national government remains the key fiscal flow through which municipalities without large income-generating mechanisms receive finance to deliver the various responsibilities that have been decentralized over the last few decades (Berrisford, Cirolia, and Palmer 2018). These financial flows from national government are rarely enough to sustain current operations of infrastructure, let alone the investment required to expand urban systems to address rapid urban growth and economic development. UN-Habitat (2015, 4) has highlighted "the near-impossible task of funding the infrastructure and services required to meet the basic needs of their growing urban population, while forward-looking capital investments are not possible for financial reasons." If even the bigger cities experience fractured fiscal authority (Cirolia 2020), this is further amplified in secondary techno-environments.

Difficulties in initiating new investments into infrastructure by these municipalities open up possibilities for outside intervention from both within and beyond the national state, questioning whether towns and small cities have any sort of autonomy over shaping their own infrastructural futures. The result is as Susana Neves Alves (2021, 248) shows in relation to Bafatá in Guinea-Bissau: "complex and shifting relations between localised state and non-state actors and the occasional flows of policies, models and resources mobilised by and with international organisations."

It is also important to emphasize that these towns and small cities act as spaces of translation between the extended rural hinterlands and national or international networks of circulation, trade, and exchange. These rural–urban interfaces suggest particular issues in the delivery of infrastructural services because of the low-density, village-type settlements that exist across many small cities and towns, particularly on their peripheries. Secondary techno-environments are also configured through their function as market towns in which daytime populations surge as farmers and traders bring in agricultural produce and find imported goods to take back to villages. This difference between daytime and nighttime populations is clearly visible in the market towns of East Africa and adds considerably to the pressures for infrastructural catch-up, again highlighting the restricted fiscal context as users of these urban spaces are often not part of the rate-paying populations that dwell within municipal boundaries.

BROKEN PROMISES, FRACTURED AUTONOMY

We thought we had made a breakthrough with our waste-management.
—"Claire," a health worker, Mbale

Planners expected Mbale's population to double from 70,000 to more than 150,000 between 2002 and 2022 (Mbale Municipality 2010). Some in the town claimed it historically had a reputation as one of the cleanest in East Africa. Yet, nearly 40,000 of its population are now living in informal settlements without adequate infrastructure services needed for everyday social reproduction (ACTogether 2015). The municipality, like other towns and small cities, has limited financial resources, with its 2021 budget less than USD 14 million, making it difficult to operate essential urban services. As "Doreen," the head of finance in the municipality, told me, "We get our grants from the center, but money is not enough. Local revenue is supposed to go on infrastructure maintenance like roads, but we just cannot generate enough." And this lack of finance included responsibility for waste-management. The head of environment, "Ann," explained the problems with local government decentralization in Uganda: "[It] comes back to money. We just don't have any. We know what we want to do, but central government just doesn't provide enough. We don't get grant funding for waste-management, so it all has to be locally raised, but we hardly get any. The environment is not a priority, and we have no grants. Local government just doesn't have the capacity to raise enough. Decentralization left us with the mandate for waste-management but not the resources. This is a false economy."

Securing necessary finance in urban Africa is rarely straightforward. The delivery of local services requires municipalities to connect to financial flows at the national scale, themselves often reliant on global institutions such as the World Bank (Obeng-Odoom 2013b). Ongoing decentralization for urban authorities, including across Ugandan towns (Saito 2012) has given municipalities responsibility for infrastructure services. But this has come without the local tax base or revenue support from national government to sustain current operations, let alone expand to cover growing urban populations (Muhumuza 2008; Okot-Okumu and Nyenje 2011). These fiscal uncertainties have meant that the municipality in Mbale is constantly playing infrastructural catch-up. These conditions mean policy makers and politicians have become open to any new opportunities to secure additional flows of investment, even if that means having little autonomy to shape plans.

NEW HOPE

It is this restricted fiscal context through which Mbale became enrolled in the Uganda Municipal Waste to Compost Program, which was primarily conceived to "reduce emissions of CO_2 by recovering the organic matter from municipal solid waste as compost and avoiding methane emission" (AENOR 2010, 5). Besides the central aim of the funders (of reducing emissions), Ugandan municipalities would benefit from upgraded waste-management infrastructures.

The World Bank was the key actor involved in its inception as part of their ongoing agenda to develop new, low-carbon, financed transformation across urban regions (World Bank 2010). They established it through a program of activities that was a "facility under the Clean Development Mechanism of the Kyoto Protocol, the world's main carbon credit scheme" (Climate Focus 2011, 8). Decision making for this investment did not come about through a democratic or open process of urban planning by the municipality. Rather, it was imposed through a powerful transnational actor that led to accusations of "throwing their weight around" by Bill Farmer, the chair of the Uganda Bureau of Carbon. In terms of the ability to address the infrastructural catch-up and shape its own future, the municipality was sidelined. As "David," the elected speaker of the municipality, explained to me and Sydney about the design of the project, "They came up with plan and we provided land; we didn't participate." The secretary of works, "Dennis," supported this contention and questioned whether the project was the best solution for waste-management in the town, asserting, "If we were given the opportunity, we would have prioritised investment in landfill."

The NEMA was tasked with coordination and eight municipalities with implementation, with the partners responsible for operating the new waste systems. The Ugandan government financed the project via a World Bank loan that provided the necessary USD 421,000 (AENOR 2010; UNFCCC 2013). The loan was to be paid back through municipal revenues accrued through selling carbon credits, with expected revenues of up to USD 26,000 annually. However, a NEMA official later reflected in 2016 that this was not to be forthcoming, and "though we have finished the verification process, we can't be sure how much and when the money will start to flow." Municipalities faced others costs of around USD 85,000 for implementation and responsibility for ongoing operation and management. Included in this local investment by the municipality was the restructuring of the existing waste facility, a collection vehicle, and twenty-eight new dumpsters across the town. The new waste-collection process would facilitate up to 70 tonnes of garbage per week being collected from across Mbale and transformed through an aerobic composting process, curtailing emissions of methane.

Once contracts and agreements had been signed by all the parties, preparatory works in Mbale and the other municipalities proceeded. They deemed it as necessary for preparing the waste-processing facilities that the waste pickers would have to be moved. Up to twenty-five waste pickers and another fifty working on related recycling nearby had to be moved from

7.3   Workers demonstrating the waste-management processing in Mbale.

7.4   The waste-management site in Mbale.

the land before construction could proceed, as reported by multiple sources, including the "resettlement plan" (NEMA 2012). "Joseph," the councillor of Doko Cell, when asked about this displacement, explained that the "scavengers were using the dump site, but there was an armed guard stationed there, so twenty or so were scared off." The municipality had used threats of violence to move the waste pickers, who for many years had undertaken the important if undervalued labor of sorting through the town's garbage. The planning process was also supposed to include community consultation, with the validation report stating, "local communities have been consulted and have demonstrated their support for the development of the programme" (AENOR 2010, 28), but no one we spoke to had any experience of this occurring. Promises of local benefits such as "the construction of a school, latrine pits, health centers and roads; the provision of scholastic materials, energy saving stoves for households" (AENOR, 2010, 28) were simply never delivered.

START UP, BREAK DOWN

Operations in Mbale began in 2009. There was a collection team in the vehicle and twenty-three workers on site for a very low wage of UGX 40,000 (USD 11) a month. Almost immediately, things did not go to plan. There was a shortage of adequate equipment to manage the higher-than-expected levels of waste. The breakdown of the only collection vehicle became a regular problem, as did, most damagingly, the ongoing relationship with a private operator. This incorporated a whole raft of issues, from the tendering process to disagreements on when operation would begin to regular payment disputes. It had forced the municipality to change contractors twice in the first few years because of a failure to fulfil the terms of the agreement. Difficulties in holding the new private operator to account remain a pressing issue. As "Doreen" contended, "The contractors and the management of this contract is one of many challenges. He takes the money, but when you look at what the outputs are, [they are] just not there."

During our time undertaking the research, the collection of waste and the compost processing had again ceased, this time because of the wildcat strike by the workers. The labor conditions were described by a frustrated local community leader simply as "terrible." Such poor wages meant mainly marginalized and elderly people were undertaking the work. Failing to pay the workers during parts of 2012 and 2013 and into 2014 and 2015 had resulted in a constant crisis and subsequent withdrawal of labor in desperate attempts to get the contractor to pay up. Operations were being disrupted, not just through technological malfunction and breakdown or lack of anticipated revenues of the carbon savings, but also through the self-organization and withdrawal of labor by the workers. This was primarily because of the municipality not being able to keep the operation in-house due to World Bank terms predicated nearly always on involvement of the private sector. Another private operator started in early 2016 after collective action of the workers, alongside growing frustration at the municipality, had been successful.

The replacement of the contractor led to a significant rise in the wage of the workers to UGX 70,000 (USD 20) a month, the issuing of personal protection equipment for use during hazardous processing, and even the promised tea and sugar for refreshments. The waste system was again partly functioning but still struggling to keep up with the amount of garbage generated in Mbale, showing this solution was itself not comprehensive enough to address the waste-management demands while generating new problems for the waste pickers, the laborers, municipality, and the town itself. There was no sign of the promised "carbon money" or of selling significant amounts of compost to local farmers. It meant the municipality was still struggling to generate sufficient finances required for everyday operation and maintenance and faced constant problems caused by the operation being run by a private-sector partner. By the time the town emerged from the COVID-19 pandemic operations had been fully abandoned, the site was no longer accessible as piles of garbage blocked entry, the trucks had broken down and the steel roof of the compost processing facility stolen. This was a story of a waste(d) infrastructure.

UNCERTAIN FUTURES

We have plenty of financial shortfalls.
—"Derick," Head of Energy, Polokwane Municipality

The inability of municipalities to address the infrastructural catch-up has fueled much of the popular discontent against the ANC-led government in South Africa. The rise of political parties such as the EFF and the

7.5 Polokwane, a secondary city in South Africa.

near-constant service delivery protests such as those taken part in by residents in Seshego in 2018 has accompanied this anger. Polokwane, the largest city in Limpopo province and the center of administration, finance, and the service industry, faces multiple urban uncertainties in navigating attempts to deliver and sustain infrastructure operations in this secondary techno-environment.

The city is booming. Ongoing rural-to-urban migration patterns seen across the country in the postapartheid era, as the hated Pass Laws were retracted, have meant surrounding populations continue to seek economic opportunities in nearby urban areas. In Polokwane, this migration is predicated on rural populations from not just within the province but also surrounding countries, including Mozambique and Zimbabwe. Polokwane's population has nearly quadrupled between 2000 and 2020, from 136,000 to 463,000 people (UN-DESA 2019). This growth has been sped up as Polokwane has become the urban center of a natural resource boom through its role as the logistical hub of these operations and as mining activities shift from the North West province toward Limpopo. Already the province is home to some of South Africa's largest gold and copper mines (De Beers, Venetia, and Palabora) and platinum, with Anglo Platinum's Mogalakwena mine, 50 km outside the city, and with estimated reserves of estimated reserves of 264.9 million ounces. Like many smaller cities, Polokwane's population and resource booms intersect, creating new challenges for under-resourced municipalities.

This rapid growth of Polokwane presents a series of uncertainties tied around fiscal presents and futures for the municipality and particularly pertaining to the operation of infrastructure services. There is a shortage in housing for low-income groups across both formal and informal settlements, meaning a need for new construction and associated services such as water and power. And with the city, incorporating large rural areas (nearly 60 percent) within its municipal boundaries, new infrastructure deployment becomes even more expensive, given the low density of these settlements. Unlike the larger metropolitan areas in South Africa such as Cape Town, eThekwini, and Gauteng, this large peripheral/rural population means that there is a much smaller proportion of urban wage earners able to pay for urban service provision through cross-subsidization—again a feature common to many towns and small cities across the region. This rural element is further exacerbated by the large townships, such as Seshego, to which Black people had historically been expelled and which lies nearly 10 km from the city, and the economic opportunities these townships offer, adding further pressures for the municipality to navigate.

CURRENTS AND PRESSURES

If the urban growth taking place across Polokwane prompts the continued need to secure finance for the infrastructural catch-up, it is worth thinking through what this means in relation to one sector: electricity. First is the challenge of expanding urban electricity networks to the 30,000 households without a formal electrical connection (Sustainable Energy Africa 2020). As "Derick," the head of energy at the municipality told me, "As the city grows, you must respond to imperatives." How to achieve this infrastructural catch-up remains the great uncertainty for municipal policy makers. With widespread informal (and illegal) electrification in many of these settlements, the ability to expect demand, ensure the upkeep of infrastructure, and adequately plan is further curtailed.

Alongside delivery of new connections and networks, Polokwane faces an equally challenging pressure to operate electricity services around nonpayment of utility bills. Many households in the city struggle to pay utility bills, as across South African cities, meaning the municipality can rarely generate projected revenues and is left with a financial shortfall. By the end of 2017, the municipality was owed ZAR 800 million (USD 50 million), with households accounting for 71 percent of these unpaid bills (Viljoen 2021). And the pressures of household debt for municipalities are not just financial. Such difficulties have provided fertile ground for political parties such as the EFF to sow discontent against politicians and policy makers. For instance, in 2016, the party marched to municipal offices with residents from areas such as Seshego to hand in a letter that demanded "The municipality must cancel all debts on water and electricity for pensioners, child-headed families and the unemployed" (African News Agency 2016).

Beyond tariff increases, which were nearly 15 percent in 2021 for electricity (City of Polokwane 2021), largely because of suppliers rather than the municipality, there have been a range of local strategies that have attempted to cover the fiscal gap caused by nonpayments. Perhaps most controversial was the introduction of PPMs for electricity. As part of the city's wider economic strategy, "Polokwane 2030: Economic Growth and Development Plan," discourses of the smart city have permeated a range of visions for the future operations of infrastructure in Polokwane. This plan aimed to adopt a "phased approach to moving towards a Smart city" through several different actions. A key imperative for Polokwane's vision concerned cost recovery for the large nontechnical losses sustained by nonpayment of electricity and other infrastructure services that have compounded fiscal pressures. The municipality's actions were particularly focused on implementing smart metering through a PPP that would use GIS technologies to mediate/measure/charge for flows of electricity, instant monitoring of tampering, and detailed location-based information for modeling/analysis. "Roger", a senior advisor to ESKOM, the national electricity utility, told me and Melu, "Nontechnical losses are so big" and that the smart meters "would improve revenues," suggesting they "could be the future of the city."

Rollout of the smart meters was rapid, with 40 percent coverage expected within three years of initial implementation (including 55,000 on

PPMs and 30,000 postpaid household meters, and 15,000 industrial locations). Terms and conditions for the PPP were signed, with the emphasis on the private partner to finance the upfront costs of installation, paid for (including extra interest charges) by future revenues (expected to be around 4–5 percent of revenue). The City of Polokwane hoped that that nonpayment would drop significantly from the rate of around 12 percent.

Using this new infrastructural "solution" came at a cost for many of the city's residents, demonstrating the ways in which this technology has harmed the urban poor throughout Africa. As "Hanef," an NGO worker in Cape Town told me, "The person with the PPM pays the highest rates per month per kilowatt hour. It is a simple equation. It's not secret knowledge, and besides, people on PPM are subsidizing because they are paying upfront for a service they have not had, whereas the wealthy get a bill after thirty days and another thirty days to pay it so this is effectively sixty days of free credit." Introducing the PPM meant it acted as a mediating technology or barrier to flows of energy for those households unable to afford the upfront electricity credit, compounding already precarious lives and struggles for survival. If the amount of credit required to access this resource flow is low, the required electricity needed to heat houses in the city in the winter, when temperatures can drop to 40°F, remained unattainable to many households in the area.

The inability to keep warm means increasing health problems in damp homes. As Cramm et al. (2011, 142) explained, "TB spreads easily in damp and crowded conditions, which are common among township households. Water incursion from internal (e.g. leaking pipes) or external (e.g. rainwater) sources causes dampness, which becomes problematic when a leaking roof causes structural materials (e.g. walls, ceiling) to become wet for extended periods of time." Hospitals treated 18,000 people in the province for tuberculosis in 2008 (Mabunda et al. 2014), a disease that is prevalent across South Africa. Tuberculosis is associated with a range of wider financial implications for the city, including the direct treatment costs that range from ZAR 400 (USD 26) to more than ZAR 24,000 (USD 1,500) for drug-resistant tuberculosis, secondary treatment costs (e.g., transport), and, significantly, the long-term loss of economic activities and associated loss of livelihoods for households. Tuberculosis is just one potential health outcome of the inability to keep warm in the winter, with problems including pneumonia (especially for HIV-positive people), influenza, and common colds, all of which have wider costs for cities such as Polokwane. Introducing PPMs meant these problems would be amplified, and it would be the poorest in the city facing the burden.

Despite the anger at the PPM, politicians followed this attempt at cost recovery by switching water from postpaid to prepaid from mid-2018. This measure was justified through fears from the municipality and utility operators about the ability to provide a continuous water supply to the city. Water-scarcity had become another pressure facing the municipality in its capacity to keep operating its infrastructure, let alone achieve the catch-up. "Denise," the head of local economic development at the municipality, told me, "We have a very serious challenge with this, importing from other municipalities in the province. We don't have enough sources of water in our district. Growth will create further stress—we need more water." They have attributed this scarcity crisis to a series of challenges, including the ageing nature of the city's water infrastructure and lack of investment in maintenance and repair. In early 2021, Lepelle Northern Water (LNW) could not supply bulk water from its Ebenezer and Olifantspoort schemes to local reservoirs, which became empty and led to a lack of water in the city and need for water tankers to undertake supply. The crisis was further compounded by the effects of drought in the province, with Ebenezer Dam suffering severe shortages for sustained periods of time. As the Day Zero crisis in Cape Town made clear in South Africa, municipalities are being put on notice that climate change is already impacting water-scarcity issues in the country (Millington and Scheba 2021). A final factor in the water-scarcity crisis in recent years has been the competing demands of the mineral-energy complex in the province. "Derick" set out the repercussions: "The worst-case scenario is when we don't have enough water. It is very scarce. The mines are already fully booking the water; most of the mines are lying close by. We have to negotiate for these resources, and they will decide."

There has been widespread anger by residents about the response to this crisis being centered on the rising costs of water and use of PPM. Water tariffs increased in 2021 by another 8 percent (City of Polokwane 2021). These unaffordable increases fed into the general mistrust about implementing water meters. A trader in Seshego told the Independent Online,

"The ANC expects us to vote into power while we don't have water and we are hungry. How do they expect us to vote for them while the billing system is wrong?" (Nkosi, 2021, n.d.). It is no wonder that this anger has found expressions in support for populist political parties, such as the EFF.

Further fiscal pressures that produce uncertainty about the financial capacity of the municipality to achieve the infrastructural catch-up or even sustain current infrastructure operations in Polokwane have included the huge 48,000-capacity stadium built for the World Cup in 2010, which created a series of problems and challenges for the municipality. While it was one of the more successful of the nation's stadiums in the post–World Cup period, it suffered, like all the constructed stadiums, from issues of usage and maintenance. Its immediate impact on Polokwane was significant, with a three-year period in which the municipality cut back on its budget to pay for a series of costs and unanticipated expenditures. As "Derick" suggested, "After 2010, all the metros that built the stadiums suffered. We got a national-level grant and a grant from FIFA, and then somehow the things we have to do ourselves have spiraled."

The municipality took out a loan (with interest) to cover these costs and implemented a "Turn Around" strategy to rebalance its accounts. The ongoing maintenance budget of ZAR 5 million (USD 330,000), together with other financial burdens, meant that the City of Polokwane was continuing to bear the costs for the infrastructure required for hosting this global mega event, with funds having to be found from the municipal budget, as the stadium cannot generate sufficient income.

At present, with a total budget in the 2020–2021 year of ZAR 4.8 billion (USD 300 million), the municipality cannot generate enough local revenue to fund the expansion of services, pay utility providers, cover the gap from nonpayment of household bills, or adequately address the water-scarcity crisis. This is despite desperate infrastructural measures such as introducing PPM, which have hurt the poorest. As in Uganda, the national government is supposed to provide a subsidy to urban areas such as Polokwane in order to overcome these municipal financing issues. Yet, the city continues to have operational issues across services such as electricity and water, as this national financing is not always forthcoming or sufficient.

Polokwane faces an uncertain infrastructural future. "Charles," a municipal official in the energy department was clear about how these infrastructural issues were already creating uncertainties about the capacity of the city to meet demands of a growing population and economy by suggesting, "We had to slow down development due to water shortages." Attempts to address fiscal uncertainties in order to generate finance for the infrastructural catch-up merely predicated another set of uncertainties that spilled down to some of the city's poorest residents. These connected and compounding infrastructural issues continue to fuel widespread public anger against the state and show the real challenges being faced by smaller cities.

OFFSTAGE POLITICS

Failing to get to grips with the experiences of towns and small cities and the secondary techno-environments that municipalities are navigating means being unable to account for the operations and geographies of infrastructure shaping the lives of tens of millions of urban dwellers. These secondary techno-environments generate a distinct set of coordinates, geographies, social relations, technologies, financial and political forces, and operations within the Infrastructural South. The futures of Mbale, Polokwane, and other towns and small cities are as important to Africa's urbanization trajectories as that of the large cities that remain at the forefront of representations, knowledge production, and urban policy formulation. Thinking about and through the Infrastructural South, as both condition and epistemology, cannot proceed from the vantage point of only the large metropolitan region but also those towns and small cities "off the map" of current debates. What happens on the supposed "periphery" may prefigure broader urban futures (both within and exceeding geographical containers, such as the South or large metropolitan regions of Africa). Mbale, Polokwane, and the elsewheres of global urbanization, far from the so-called modern cities of the metropole and the contemporary, iconic megacities of the South, are integral to urbanization trajectories, the making of urban life, and urban futures.

In thinking through these secondary techno-environments, I contend there are some clear politics and imperatives that come into view. First, questions of urban autonomy are never far from the surface, especially pertaining to the capacity to address the infrastructural catch-up. This autonomy (or the lack of it) to govern and plan without undue interference from the national state and institutional institutions has both fiscal and political dimensions. It makes up a profound democratic gap for the politicians, policy makers, and communities of towns and small cities.

In Mbale, the design of the waste project was supposed to address local imperatives, but these were simply not being addressed. "Peter," a municipal official was very clear: "This is a World Bank project with its own set of guidelines; they have their own ways." Under the fiscal conditions that mediate secondary techno-environments, actors such as the World Bank can impose a vision of infrastructure operations while also proclaiming, "The entire program is voluntary in nature." Such a capacity to intervene in these techno-environments means the institutional concerns of actors such as the World Bank come to take precedence over the people and civic institutions of smaller cities and towns such as Mbale. These processes are also connected to much longer histories of structural adjustment and colonial rule in the area. When I asked "Robert," the elected secretary of works in Mbale about the future of the town, he envisioned a fiscally "autonomous city" to "accommodate the immense needs of the population and commercial growth." Clearly, demands for urban autonomy are a critical politics for the smaller cities and towns across the Infrastructural South in allowing them to experiment with new ideas, devolve power, open up local economies, and incorporate the wishes, knowledge, and demands of communities.

The struggle for increased autonomy has often been tied to demands for recognition by towns such as Mbale for city status and the extra financial resources that this can bring fiscally and to be deployed in infrastructural catch-up. As "Ruth," the head of community development in Mbale, told me, being a city meant that "you get more financial support than if you are a town." In Mbale, this claim for city status was decades long, pleading with the reluctant national government that was finally achieved in 2020. However, as Mr. Wegulo, the LC1 chairperson of Cathedral cell, told the Monitor newspaper, "It is still a city on paper. Nothing has come with it so far, not even in terms of cleanliness. Heaps of garbage are everywhere" (Wambede and Masongole 2021).

Second, given the lack of urban autonomy, failing to achieve the infrastructural catch-up generates popular discontent and anger among communities in towns and small cities that seem likely to remain part and parcel of everyday urban life. As I showed in Polokwane, the rise of populist parties such as the EFF in the postapartheid years and waves of protests by residents around service delivery issues emerge from the deep frustrations and struggles for survival faced by communities. The promise of a better future does not always materialize, and as the case of introducing PPMs showed in Polokwane for some households, attempts to manage infrastructure through fiscal and other types of crises can generate further anger. This is often a messy politics, one in which groups, associations, and communities exist outside the domain of "civil society" and use various tactics and strategies such as protest to assert claims for the infrastructural catch-up. This politics highlight how the domain of infrastructure is so important in the mediating of social relations across smaller cities and towns. In Mbale, discontent from striking workers emerged out of the privatized contract, leading to wildcat strikes that, while extended, did ultimately succeed, at least temporally improving conditions. The failure of the waste-management process generated anger in surrounding communities, with the named mounds of garbage perhaps the most humorous response. Mbale's residents vented their deep frustrations at the inability of the municipality to achieve the infrastructural catch-up.

In Polokwane and in Mbale, various forms of uncertainty about the infrastructural future pervade the operations, governance, and everyday experience across these secondary techno-environments, from water-scarcity issues in Polokwane to waste-management in Mbale through to worries by households about whether energy flows or workers who are unsure when they will be paid. This is fiscal and political turbulence that reverberates across these urban spaces in which uncertainty over when or even whether the infrastructural catch-up can be achieved is always present. And there can be no end to this uncertainty. An official in Polokwane told me in relation to the water crisis, "We are thinking of building reservoirs now" in response to questions about how to navigate the concerns about scarcity and its impact on service provision. And yet, as I have argued with colleagues, "even when solutions to uncertainty are found, there is hardly ever a final resolution or fixed point of closure . . . uncertainty is rarely, if ever, eradicated from the urban milieu; rather, it is managed, displaced, deferred, reconfigured, or reproduced" (Zeiderman et al. 2015, 299). Clearly, a techno-fix such as new reservoirs would simply generate new uncertainties, much like introducing electricity and water meters did.

If this uncertainty about the infrastructural futures of towns and small cities across urban Africa is imbued with the inequalities generated out of failure to address the infrastructural catch-up, there is also the possibility that such uncertainties may lead not to ongoing crisis and difficulties but perhaps to somewhere else. Some commenters and policy makers have suggested small cities and towns could become important alternatives to larger cities for sustainable urbanization, as "compared to large cities, secondary cities and towns play pervasive economic roles as vehicles for rural–urban transformation and poverty reduction" (Agergaard, Kirkegaard, and Birch-Thomsen 2021, 16). Discourses that swirl around such claims revolve around the notion that the large cities have become ungovernable and unplannable, and therefore increasingly difficult to deliver necessary infrastructure. For instance, the World Economic Forum (2019) promotes the idea that "Africa's economic future lies in its smaller cities," even as it sets out "that when it comes to infrastructure, financial investments, land development, and urban management, few African secondary cities have the autonomy to pursue long-term strategic planning." However, as I have demonstrated in this chapter, current fiscal conditions and the uncertainties about how to navigate the demands of the infrastructural catch-up mean the potential of these urban spaces remains unfulfilled and unlikely to change under current political-economic conditions.

8

DIGITAL DISRUPTIONS FROM "ABOVE" AND "BELOW"

THE DIGITAL PRESENT

In neighborhoods across East Africa, a popular economy operates to service, maintain, and repair the vital infrastructure that allows urban dwellers to access various digital interfaces: the phone. This is because "the smart city at the margins is dominated by mobile phone access" (Odendaal 2018, 262). A walk around cities such as Nairobi or Kampala shows the importance of these forms of maintenance and repair to the continued viability of this technology to operate. This includes charging, mending, or repurposing handsets, putting credit on them, Mobile Money transfers, upgrading software, and buying/selling (maybe acquiring in other ways) various components or the units themselves. It is in these various forms of maintenance and repair that sustain access to the digital realm. We can think of the people, things, practices, and technologies involved as a "people as infrastructure" (Simone 2004b) for the digital age.

The popular economy that has built up around the maintenance and repair of mobile phones and the capacity of this technology to access various types of interfaces offers possibilities and potential for people excluded from formal job markets. A self-described fixer, "Peter," in central Kampala, explained how selling credit "provides business opportunities to youths who have not completed education." "Peter" suggested that this was

8.1 Fixing phones in central Kampala.

because "it doesn't require serious training, only a little guidance, and you can run it. For Mobile Money, the providers call you to train you for a few hours, not the whole day; the training only occurs when they are issuing new SIM cards for Mobile Money."

This young man's particular role in the maintenance and repair of phones was more highly skilled than some of his peers and involved learning from more experienced hands over several years, as well as continued experimentation, tinkering, and (re)making. As we watched, we saw him inserting and breaking up circuit boards, SIM cards, handsets, and batteries. In effect, "Peter" built his livelihood on the demand for tools to enable and sustain access to the emergent digital techno-environments of the Infrastructural

8.2 Phone charging kiosk in Namuwongo.

South. He outlined his aspirations, suggesting, "My business will grow to be a big company, using Mobile Money to upload airtime and pay bills, like water, electricity, and many things." In his vision of the digital urban future, this fixer expected to gain new skills to perform tasks regarding the servicing of phones. The aim was to diversify his labor into the multiplicities of opportunities being actualized through daily use of mobile technology in the city and the new maintenance and repair regimes being established through the digital interface.

Digital openings have come into view, especially for young people excluded from formal labor markets,[1] offering them new openings and promises to take part in the urban economy. As this chapter will show, these take multiple forms, from Uber taxi drivers using their smartphones to pick up customers to tech-savvy entrepreneurs attempting to solve long-standing urban infrastructural problems, such as congestion, or youths working in a kiosk and ensuring that phone users remain plugged into the network. These digital processes and spaces enhance ways to become involved in the economy, configure livelihood opportunities, and open up expectations of improvement (Di Nunzio 2019). These openings prompt hopes of consolidation amid precarious situations. But the digital present is also proceeding in broader political economic forces in which this online space is generating new top-down forms of control and extraction that portend a future more in line with the techno-dystopias conveyed in sci-fi films such as *Moons over the Monsoon*. Clearly, the capacity of digital techno-environments to transform urban conditions is already proceeding, but the futures that emerge from this digital present remain to be fully coded.

DIGITAL GEOGRAPHIES AND OPERATIONS

From the rise of tech clusters, such as Silicon Savannah in Nairobi and Yaba's transformation into Silicon Lagoon in Lagos, to the upheaval caused to the banking sector by Mobile Money services, such as M-Pesa, now used by upward of 37 million customers (Medici 2019). The ever-growing assortment of new mobile phone apps spans food delivery services, such as HelloFood, the Ghanaian mobile games service, Leti Arts, and the South African medical support initiative, Vula, connecting health workers to specialists. Everyday urban life is now increasingly integrated into the apps, algorithms, programs, fiber-optic cables, monitoring systems, devices, and interfaces that incorporate these emerging techno-environments, offering possibilities and problems alike. Digital mediation is a powerful tool to intervene in, configure, enhance, and potentially contest the city, even as the state and corporate sector deploy new forms of control and extraction. Much of the existing, let alone the yet-to-be-built, infrastructure in the region will probably be operated, experienced, or monitored through code and software. The digital is disrupting operations and geographies of infrastructure in a myriad of forms. To think of the Infrastructural South today means to engage with this digital present, the technologies of African "smart" cities, and the ideas, experiences, and innovations of people navigating these disruptive waves. It also means to speculate about futures that are only just beginning and best imagined through African sci-fi novels such as Tade Thompson's (2019) Wormwood Trilogy, full of characters such as Jack attempting to reprogram a sex robot for some future insurrectionary mission in 2050s Nigeria. Thompson (2019, 2013) writes about Jack and how "he spends days accessing books on Artificial Intelligence.

He cannot code, but it gives him a good idea about the specifics of what he needs to tell the person he hires. He thinks the bot has a kind face." The passage highlights the ways in which code and robotics have become part of life and the ways in which protagonists such as Jack attempt to hack such configurations for their own purposes.

Nascent techno-environments of the digital are an ever-shifting mutation of the universalized technological model of what came before. Indeed, a wave of disruptive technologies places urban Africa at the global forefront of the so-called digital revolution. But how do we comprehend the operations and geographies of these transformations? Existing literature on smart cities has produced "global" insights and theorizations, predicated on experiences of cities in the North. It means, like traditions of urban theory, which this book has critiqued, that explanatory frameworks are rooted in urban contexts that have differentiated histories and geographies of infrastructural modernity: from long-standing, universal systems to minimal housing informality to concentrations of large-scale ICT companies and high employment levels.

The smart cities of these regions that we term "the North" form a divergent digital landscape from that of urban Africa. We must therefore be attentive to such a difference and shift beyond abstract conceptions of space configured through code and software. Rarely do actual existing processes of flawless techno-managerialism become clear. Indeed, despite promises of digital technologies acting "as a modernizing tool, governing force and precipitant for curing urban problems, boosting urban development and entrepreneurialism" (Guma and Monstadt 2021, 360, deployment encounters the messy techno-environmental conditions of cities.

We should challenge understandings of how urban Africa is playing catch-up and facing a deficit across a "digital divide" (Fuchs and Horak 2008). Amid the accelerating deployment of the digital in celebrated tech-savvy hotspots, such as Nairobi and Kigali, linear and Western-centered notions of technological progression and development require disruption. The shaping of urban space through the digital has become ubiquitous and integrated into everyday life through the rapid transformation of the networks and user practices of urban Africa into the twenty-first century. According to the Disrupting Africa: Riding the Wave of the Digital Revolution report by the consultancy PricewaterhouseCoopers (n.d.), between

2007 and 2016, the growth of mobile phone usage jumped 344 percent from 174 to 772 million people. This rapid uptake now accounts for well over half of the 1.2 billion people living on the continent.

Understanding how digital disruption is unfolding should begin with the acknowledgment of the centrality of "mobile technologies in shaping new sociotechnical constellations" (Guma 2019, 4). The rapid mobile broadband growth rate has accompanied this phone-led transformation. Between 2015 and 2016 alone, this increased by 58 percent (PricewaterhouseCoopers n.d.), allowing tens of millions of urban dwellers to access apps, social networks, and the possibilities of digital techno-environments. This does not mean ignoring those who cannot access these services—indeed, "ICT access correlates with higher incomes and clustered private investment, thereby perpetuating inequalities" (Odendaal 2011, 2394)—or the context within which mobile users in Africa in 2019 were paying nearly three times the global average for voice calls and the internet (Kazeem 2020). Rather, such statistics convey the rapid uptake of mobile phone technology and access to digital worlds, often within spaces in which infrastructure is partly absent or being newly installed. The PricewaterhouseCoopers report highlights the supposed potential encased within this rollout: "One of the great advantages Africa has over other continents in riding the disruptive wave is that there's far less legacy to get in the way than in other regions, creating a clean sheet on which companies can develop their own distinctive business models."

Boosterist narratives of digital revolution in Africa echo previous eras of optimism around the capacities of ICT to enable and speed up so-called development through various types of leapfrogging of existing global technology regimes, which have anchored urban life in the north (Alzouma 2005; Amankwah-Amoah 2015), with the most often-used example being that of the jump from landlines to mobiles over the last two decades. Leapfrogging has transformed how people use all types of infrastructure. For instance, in 2003 in rural eastern Uganda, it took the NGO for whom I was working months of pleading, harrying, hustling, and endless waiting at the telecommunication company offices to get a landline installed, joining only 27 million users registered across the continent (Meldrum 2004). The landline provided an essential service, most notably through allowing the rapid delivery of emergency drugs from the nearby town to the health clinic, improving treatment of patients. In doing so, this ICT network hastened the circulation of various components: medical expertise, life-saving drugs, blood, and communication between different health professionals. For many, the option of a landline telephone remained unfeasible in the following years. Today, access to mobile technology means the benefits of telephone access (and the digital realm) are exponentially greater than two decades ago. And yet proponents of the techno-utopianism of leapfrogging ignore some important factors in that they "share a set of dispositions and worldviews which are highly 'modernist' and technocentric, characterized by a propensity to view and act in favor of exogenous 'technological' solutions to development problems" (Alzouma 2005, 339).

In recent years, digital technologies and accompanying smart city visions have also become integrated into corporate and state-led strategies for urban development (Watson 2014; Odendaal 2016), as well as across the new cities and urban extensions already examined in chapter 3. The allure of digital technology has become integral to future imaginaries of a modern urban Africa, despite suspicion concerning promises from various actors, including development agencies, social movements, and municipalities. Global and regional governance initiatives, such as the African Union's Agenda 2063 and the United Nations 2030 Sustainable Developmental Agenda, have mirrored and drawn from the mobilization of smart city strategies and the emerging role of the digital in urban development (Slavova and Okwechime 2016; Pieterse, Parnell, and Haysom 2018).

Advocates of the digital orientate discourses around various benefits, including enhancement of infrastructure to expedite circulation of various types, economic growth, addressing poverty and environmental challenges, and new livelihood opportunities. Digital techno-environments are simultaneously an envisioned future and, increasingly, an everyday experience for tens of millions of urban dwellers who, like elsewhere, have learnt to navigate these thick flows of code and software, sensor network data, and constant streams of analytics. I am interested here in the intersection between the growing visibility of the digital in broader narratives on urban development, the construction of particular accumulation opportunities through new forms of local and global technology, and the often less-visible ways in which people experience these disruptions.

Perhaps the most celebrated example of this digitalization of everyday urban life is the wave of financial services that various companies have established through the interfaces of mobile phones. Advocates of these new digital tools, often developed in the region, argued that they have helped to address urban inequality. As Njuguna Ndung'u (2018, 37) suggested, "At the outset, mobile phone based financial services, starting with M-Pesa, were viewed as a technological platform that would allow a menu of financial services that would be offered to Kenyans and would drive financial inclusion: it has stood the test of time and has been vindicated."

The digital techno-environment is entangled across urban space, enabling new types of circulation, associational life, livelihood, and opportunity. The functioning of the digital in these spaces challenges a teleological narrating of urban modernity. The "digital divide" between Africa and Europe persists, despite the accelerated penetration of various technologies into everyday urban processes. This reflects longer histories of imperial infrastructure, uneven center–periphery relations, and the global political economy of technology. Concurrently, the digital emerges from across the innovation spaces of the South, bypassing and making obdurate long-used systems and associated infrastructures in Europe or North America, from those of high-street banking to telephone landlines. These digital disruptions are precipitating South–North technology transfers, most notably Mobile Money, while a growing flow of ideas, innovations, and experiments are emanating from urban Africa.

Attempts to decipher these geographies and operations of digital techno-environments in urban Africa are important. Like broader conceptualizations of urbanization, I contend that much of the literature on smart cities or digital infrastructure derives from experiences of paradigmatic tech cities in the urban North (Datta 2019b). This is despite the "technological advances that are emerging in the urban South, not least in African cities . . . from high-tech and smartphone-based applications to low-tech and feature-phone-based platforms" (Guma 2019, 4). And most importantly, in thinking about the Infrastructural South in specific reference to Africa's urbanization are relations between digital uptake and persistent urban informality. Unlike the (supposedly) ordered city spaces of northern tech hubs, such as San Francisco, Helsinki, or Tallinn, this integration suggests a "collusion of informally organized digital and urban networks that simultaneously disrupt and facilitate formal structures" (Janu 2017, 117). Clearly, then, urban Africa is more than a "passive recipient" (Guma and Monstadt 2020, 5) of digital technology, given both its particular techno-environments and its generative role in innovation.

DISRUPTION FROM "ABOVE"

ELECTION SHUTDOWN

Clampdowns on social media platforms and the democratic possibilities they enable are increasingly visible in some African cities, as elsewhere. In early 2019, the government of President Joseph Kabila cut internet access and text messaging in the DRC after the national elections. And in Zimbabwe, the high court ordered an end to the blocking of the internet, after a week of fuel protests that led to a severe, violent military clampdown, ruling that the minister of state for security did not have the powers to enforce such a ban. President Emmerson Mnangagwa's government relented but kept Facebook, WhatsApp, and Twitter offline for longer, highlighting how social media are shifting the capacity of the state to order circulations of people, ideas, and information during times of political turbulence. Datta and Odendaal (2019) develop Mbembe's (1992) concept of the banality of power to think through the everyday use of smartphones. They consider how "the interspersing of routine with sporadic and concentrated acts of soft power and brute force that make the smart city an embodiment of state governmentality" (388).

I was traveling into the city on the day the results of the 2016 Ugandan elections were to be announced. The streets of Kampala were unnervingly quiet. The presence of large groups of soldiers and a range of intimidating weaponry were on show on the road to Entebbe Airport. Strategic sites around the city and at the base of operations at Independence Park in Kololo hosted hundreds of soldiers. The military seemed to be camping out in the capital, awaiting any sign of contestation. While there had been no clear decree by the government aimed at keeping people in their homes, there was widespread fear and uncertainty. However, the historic behavior of the police and army across Kampala in postelection periods meant that there was little desire to be seen moving outside (even as preelection mobilization was somewhat tolerated).

The movement of people, especially opposition activists and politicians, had remained controlled and halted completely, showing how the

8.3 People traveling around the city in preelection campaigning were not allowed to move after the election.

state had moved from the all-out violence of earlier elections (particularly 2001/2006) to a strategy in which the repression is less visible and blatant just but as damaging (in different ways) to those seeking to challenge the regime.[2] This government clampdown was especially focused on the FDC in 2016 (shifting to Bobi Wine's National Unity Platform in 2021). Opposition parties had been systematically disrupted by the various arms of the state after the last voter had left the voting booth. Citing fears of public disorder, the police played the most prominent role in such restriction of movement that had felt like the slow-motion, even mundane suffocation of the opposition, rather than the visible violence unleashed in previous elections.

Attempts by the opposition to mobilize supporters on to the streets of Kampala were nearly always met by high levels of violent repression, and the state was now also focused on closing down the internet, viewing it as a potential space of dissent. National and international media heavily criticized disruption of these digital infrastructures as an attack on freedom of speech (Index for Censorship 2016). The president justified such actions by arguing that "steps must be taken for security to stop so many [social media users from] getting in trouble; it is temporary because some people use those pathways for telling lies" (Nation 2016). The state had control of both the streets and the digital realm.

The shutdown generated a whole series of secondary disruptions across the city. One of the most damaging was that the government also froze Mobile Money transfers, bringing panic to people who relied on this digital banking system. As the mobile phone companies, who control the banking, were unwilling to challenge the government or seek ways to navigate around the measures used by the Uganda Communications Commission (UCC) to curtail money transfers, it created an immediate crisis of social reproduction for many Ugandans. There had been no warning that the UCC would block this service. People up and down the country waited for money transfers for food, travel, and medicine, some were left stranded for nearly four days until late on the Sunday evening. The inability to use Mobile Money transfer compounded the unwillingness of rural producers to travel into the city, especially in the final days beforehand, creating a significant impact on prices as reported by market traders such as Ali in Namuwongo, who saw the price of potatoes surging by nearly 50 percent. Increased food prices precipitated a sharp rise in inflation by 1.8 percent to nearly 10 percent. "Hamza," a trader in the central market, explained what this meant: "I faced a hard time because I could not access money for my business since I use Mobile Money . . . so I had a big challenge. Prices of goods were hiked mostly because of transport so this led to increase in food prices brought from outside the city." This digital disruption showed the power of the state to disrupt not just political organization but also the very basis of social and economic life that had now been so integrated into the digital techno-environments of Ugandan cities.

The UCC, based in a bleak, postmodern Orwellian building in the suburb of Bugulobi, must have been full of shocked management by lunchtime on Friday as it became apparent that their tactic of disrupting digital communications had backfired. This attempt to stop all flows of information, of knowledge, and, importantly, reporting on the election had clearly not had the intended outcomes. Alongside the wide-ranging criticisms within and beyond Uganda, it was clear from the ongoing buzz across social media that many Ugandans had simply ignored the ban and navigated this attempt to limit flows of information, discussion, and data. The most popular way was through VPN, with more than 1.5 million downloads in Uganda on the Friday alone, as it became an essential tool for many parts of the (often younger, urban) population. The government and its security agencies had great success in clamping down on movement and circulation within the city. It had been less successful in imposing the same type of restrictions on flows of online information, opening up new ways for the opposition to mobilize. A FDC opposition leader in the city told me, "We were expecting the social media ban, but since communication via phones was not banned, we have used this. The VPN has been a good tool. We tried to alert people before the election, but many didn't know how to set it up. It was difficult, but next time they will know . . . We have technology experts." And it wasn't just activists who were responding to this attempt to close down platforms. "Elizabeth," a resident of Namuwongo, explained, "Many people had not known that social media were blocked until they realized other people were installing VPN to access social media like Facebook, WhatsApp, and Twitter most commonly used by the people."

The problems in Kampala during election time, as in some other African cities, from Bulawayo to Kinshasa, suggest a less optimistic digital future than the promises of consultancy reports, tech start-ups, and economic development agencies. The importance of the digital techno-environments of the city was exploited by authoritarian governance actors during moments of urban conflict and contestation, seeking to limit the potential and possibilities of new forms of digital organization and resistance.

Ugandan government actions under President Museveni seemed determined to resist the democratic currents that have emanated and reverberated through digital infrastructure. The government also sought to use legislation to limit the potential of social media, most prominently by imprisoning the inspiring Dr. Stella Nyanzi under sections 24(1) and 24(2) of Uganda's Computer Misuse Act. The introduction in 2018 of the Over-The-Top (OTT) tax for service users to access social media through payment of USD 0.35 per week to access services such as WhatsApp, Facebook, Skype, and Twitter. According to Juliet Nanfuka, speaking in the New Vision newspaper, the results of this have been catastrophic, with "at least five million Ugandan internet users going offline" As "Peter," the young man operating the phone kiosk, explained before the implementation of the Mobile Money tax, "Of course our profits will reduce because some customers will stop using Mobile Money, because most of our profits are commission based. So, if we have few customers, then that means there is going to be small base of commission." And such worries soon turned out to be prescient, with the Uganda Debt Network showing in 2019 that 35 percent of proprietors laid off workers (Naisanga, n.d.) at phone kiosks—the very sites in which new livelihood opportunities had previously been generated. Attempted coercive usage and control of digital techno-environments clearly spills out from the political to the economic and social realm.

Amid the first wave of blocked cities, Bulawayo, Kinshasa, and Kampala, this disruption of access to the internet was crude, and in the Ugandan case, people quickly took up VPN to navigate some of this digital disruption. Implementing various surveillance and tracking technologies seems likely to become increasingly attractive to undemocratic governments such as Uganda's, even as urban dwellers seek to navigate these attempts at control.

DIGITAL EXTRACTIONS

Established global platform companies involved in the transport sector seek ways to expand accumulation opportunities into new markets predicated on the commodification of thick streams of customer, routes, and driver data (Barns 2019), including through the promise of enhanced transport options. One of the most prominent of these corporations is the ride-sharing app, Uber. Valued at USD 82 billion in the run up to its initial public share offering in 2018, Uber has sought market expansion through the promise of a mobility revolution for urban Africa. Introducing this platform to Kampala in 2016 meant the city joined 462 others around the world. The firm, whose headquarters are in San Francisco, initially entered the African market through South Africa in 2014, providing 2 million rides within the first six months of 2015 (Cotterrill 2015) before expanding to seven other countries. I consider Uber a digital disruption from "above" because the company developed the app outside urban Africa and only then subsequently deployed it in cities such as Kampala. Uber built its design, software, code, and user interface through early testing, deployment, and operations in North American cities. It is in effect a US corporation seeking to open up accumulation opportunities in urban Africa via promises of enhancing mobility in order to drive its growth and shareholder value. But what costs does Uber entail for a city such as Kampala?

Uber has arguably already made life more difficult for taxi drivers in Kampala through negotiating tight margins and undertaking various types of livelihood hustle, suggesting that digital layering across the city's road network can have varied effects on existing users and operators of transport infrastructure. Joseph, who had worked as a taxi driver for more than two decades, described how he and his colleagues felt pressured to join Uber when the corporation arrived in the city, as they were worried about losing customers; 15,000 drivers signed up within a year with the service. Joseph and his fellow operators continued with business as usual from their "spot" or "stage" in Bugulobi, just outside the center of the city and close to the UCC offices. They also operated on Uber, resulting in the potential to generate work across three different spaces: the traditional local "spot" where customers can grab a taxi and drivers hang out between jobs; the long-standing network of customers they had established through the hustle of everyday taxi driving, who would call them when required; and this new digital interface via which customers could hail the drivers through a few swipes of a smartphone somewhere in the city. In effect, Uber was also promising taxi drivers enhanced and extra layers of income generation.

The experiment with this platform did not last long for Joseph. Uber demanded a 25 percent cut of all fares, while the rapid increase in taxi drivers in the city had also been pushing down the price of a journey. Alongside rising petrol prices, Joseph soon found that the operating margins were too tight. Normally profitable, longer trips to cities such as Entebbe or Jinja were generating little profit, especially if becoming caught in the ubiquitous traffic of the Kampala region. Sometimes he could only charge UGX 75,000 (USD 20) and would have to use 25 percent of the fee to pay Uber and another 20 percent for petrol. With more than 1,000 Uber drivers available in Kampala in 2019, the competition was already becoming fierce. This has affected the livelihoods of established taxi drivers such as Joseph, who have been forced to work longer hours to ensure they generate as much income as before this global corporation touched down and engineered its operations onto Kampala's road network. With Joseph and his fellow taxi drivers reverting in the main back to more traditional taxi trade practices, Uber has continued to increase its pricing. The costs for drivers reached a breaking point by late 2019 as they went on strike, storming the company's offices over pay and treatment. A striking driver told the Nile Post, "We joined the company willingly and by then, we were being referred to as partners but we have been relegated to slaves in our own country" (Kazibwe, n.d.). The costs of the promise of enhanced mobility for Kampala's residents seem to be borne by the taxi drivers of the city, with the profits captured by this US corporation. Uber is a clear example of a new form of infrastructurally enabled extraction from the West.

Uber's entry into Kampala highlights how global tech companies seek new frontiers of digitally enabled accumulation across urban space. The tech company extended this coded layer of commodification across everyday life in the city. The process has parallels with the historic natural resource extractions operating in urban hinterlands. This is because digital technologies can act as an extractive force in African cities, with Uber offering a compelling case of such logics at play in Kampala. I primarily understand this extraction through the way Uber presents itself as a tech company, not a transport company as, "the technological heightening of competition in the labor market represents an effective strategy of control" (Carmody and Fortuin 2019, 9). In effect, Uber has sought to capture the value-producing operations of taxi drivers in Kampala across its road infrastructures and to extract payment from people already working on tight margins. Digital techno-environments in this case operate on the logic of new types of data-driven information flows that are enabled and connect the urban dweller to a controlled corporate platform through a mobile phone and the promise of enhanced urban mobility. A customer taps a button, a taxi is dispatched, and the driver takes the user to a different part of the city at a potentially higher speed than would have been possible previously. However, the cost of such digital enhancement to transportation infrastructure is carried by drivers, such as Joseph, as it threatens long-established livelihoods. The operations of Uber will probably face further resistance and contestation as it seeks to dominate the taxi industry in Kampala, as it has done elsewhere.

DISRUPTION FROM "BELOW"

ENHANCED CIRCULATIONS

Layering of digital enhancements onto the road infrastructure of African cities is likely to reconfigure how urban dwellers, goods, services, and communications travel across and between these spaces. The Oxford English Dictionary definition of enhancement conveys "the act of increasing or further improving the good quality, value or status of somebody/something." The UgaBus deployed in Kampala offers an interesting example of how the digital is enabling this.

Digital disruption to existing geographies of urban mobility in Kampala are emerging beyond the extractive logics of global corporations such as Uber. I understand these processes as the digital city being configured from below. Here, I wish to emphasize how these digital intermediations onto existing infrastructure are emerging out of the everyday experiences of people who navigate urban transportation and are seeking to enhance daily practices of city life. As I set out in the introduction, urban Africa is lauded by many consultancies, media outlets, and tech leaders for digital innovation across urban space. This has included locally produced innovations targeted at transport operations and offers a counterview to the corporatist smart city operations of top-down North-to-South digitalization.

UgaBus is a basic yet interesting app that mediates flows of buses and associated socio-ecological circulation across and beyond Kampala. This start-up can help convey how the digital can intersect with existing infrastructure and mobilities to produce more efficient, enhanced circulation across and beyond a city. In a few short years, UgaBus has become the largest online interface for booking bus journeys in Uganda, helping users navigate the myriad of competing operators through making seats available online to travel across East Africa, incorporating more than 100 daily departures. Passengers access UgaBus through its web portal, pay through various means, including Mobile Money and even Bitcoin, receive an SMS notification they can take to the bus station, and get on their transport.

A young entrepreneur, Ronald Hakiza, developed UgaBus. He shows how local innovators draw on knowledge and skills in relation to digital technologies to remake the infrastructural conditions within which they are situated. Inspired by the often-bruising encounters at Kampala

8.4 An intercity bus in Uganda.

8.5 Congested streets of central Kampala.

bus station, and one particular experience in which he nearly missed a relative's funeral, Ronald attempted to make the system more efficient and user-friendly. Using an initial loan of UGX 600,000 (USD 150) from his former teacher and more funds from his aunt later on, Ronald built a demonstration app that meant he could access further investment through a grant of USD 10,000. The app gained popularity among customers, and in 2019, the company relaunched to enable it to operate with a larger user base. In 2017, Ronald valued his company at UGX 108 million (USD 30,000)—a solid achievement for this young entrepreneur.

The UgaBus venture offers a pertinent case for thinking about how new infrastructural actors become involved in digital layering and how they deploy these innovations across extant operating systems around notions of enhanced efficiency. In particular, the key logic of this app is to improve mobility across often-degraded road infrastructures. This was undertaken by enabling the accelerated circulation of people, goods, and information across and between cities. The first type of enhanced efficiency in terms of this circulation relates to the role of the app in decreasing the sometimes-chronic congestion in central areas of Kampala (Matagi 2002), which mirrors the growing gridlock across much of urban Africa. These conditions are predicated on population and vehicle growth, the ever-further peripheries of many cities, and the lack of investment in new infrastructure. In Kampala, traffic jams are especially notorious in the old downtown around the bus station. By offering users an opportunity to purchase tickets without having to go into the city, UgaBus could cut down on long, unnecessary journeys to secure a ticket.

A second type of enhanced efficiency that UgaBus produced was through shifting the intensity of circulation, which comes from ensuring that buses are full, with each passenger being allocated a seat, meaning that fewer half-full vehicles are leaving the bus station, further reducing the number of problematic vehicles circulating in the urban space. These reduced flows include the traffic on the road and the toxic pollutants that come from what are often very old buses. Here, we can see how placing a digital layer onto existing infrastructure can have material effects. A third type of efficiency relates to the potential financial savings made by passengers in being able to access real-time updates on bus-seat availability and prices. Digital technologies promise to reduce transaction costs for both customers and operators (Barns 2019). Commentators have characterized this part of the digital wave as constituting the "sharing economy" (Richardson 2015). This means that money circulates in local economies in other ways; cheaper travel and mobility are critical for household livelihood strategies, with long-standing evidence that economic barriers exist in accessing transport in African cities (Olvera, Plat, and Pochet 2013). The UgaBus app enables quicker mobility for those livelihoods predicated on the transport of goods, whether working across different cities or between cities and villages.

Fourthly, it is worth highlighting the importance of how the app typifies how digital intermediation from below can make legible the dense, overlapping, and informal systems involved in securing a bus trip in cities such as Kampala. The public face economic and social transactions of the intercity bus system. I would contend that the flattening of these dense topographies of associational life around bus operations should be understood as a kind of efficiency in itself. This is because the app becomes a tool in navigating infrastructure and the shifting socioeconomic conditions of securing a particular kind of mobility, at a precise moment and place.

The case of UgaBus highlights how new actors are producing digital techno-environments not through static technologies but rather with constant updating via the algorithms that are acquiring data and feedback from user experiences (Coletta and Kitchin 2017) or new software aimed at enhancing functionality. It is this constant state of mutation and of being "in the making" that connects the digital to the broader making of the Infrastructural South. Ronald described the need to make sure the UgaBus interface is constantly evolving: "Everyday, customers come up with new things that they think we should include. So, everyday, we are innovating; everyday we have to add in new things—and believe me, I do not think we have a complete product yet. We are innovating every single day."

Problems persist in the smooth functioning of UgaBus, demonstrating that the promise of digital enhancement might experience breakdown. This comes as disruption for users and unanticipated costs for the operators, as Rob Kitchin and Martin Dodge (2019, 47) argued, "Smart city technologies are promoted as an effective way to counter and manage uncertainty and urban risks through the effective and efficient delivery of services, yet paradoxically they create new vulnerabilities and threats." In 2017, hackers brought the UgaBus system crashing down for a month and created serious problems for customers who had already purchased tickets and for operators to decipher who had paid and which passengers were supposed to be on particular vehicles. This interruption played out in real time across multiple social and economic interactions. These techno-environments cannot always be anchored to an optimized and controlled city, even as transport systems are reprogrammed into these technologies (Barns 2019).

Apps such as UgaBus—and indeed, more broadly, digital technologies— rely on continued flows of electrical power and therefore involve anticipating, addressing, and mitigating the disruption of an urban energy network prone to constant interruption. In reflecting on these underlying vulnerabilities of a digital city, a phone fixer in Namuwongo was clear about the inability to always find an answer to address disruption: "There are no alternatives. Sometimes, people leave their phones and just wait till the power comes back." And the return of electricity after disruption creates further requirements, as it "comes back and destroys the chargers." It means that the fixer always has to anticipate, calculate, and ultimately take risks regarding the flow of electricity to their workspace in order to keep the digital city going. Like the fixers working out of the ubiquitous phone kiosks on almost every street corner, Ronald has also had to plot ways in which to navigate these underlying vulnerabilities and highly unstable spatiotemporal dynamics of digital techno-environments, be it hacking or disrupted power. It means that apps such as UgaBus cannot always fulfil their promise to enhance the operational flows and circulations of the city.

The disruption of existing infrastructural geographies through digital tools takes on multiple forms across the daily operations of the underlying technologies, hardware, and infrastructure of everyday urban life. As I set out, this includes the risks of malfunctioning interfaces and operations resulting from hacking and other malicious means, which might take a digital interface offline and predicate a secondary wave amid the everyday flows and circulations of the city. Electricity remains a critical underpinning to the operating of digital urbanism. Frequent interruptions, a thorough lack of generating capacity, broken and under-maintained infrastructure, or simply a lack of money to buy electricity credit can all delay, disrupt, or unsync the ability to access the digital interface for the urban majority.

CONTESTED REGIMES OF MAINTENANCE AND REPAIR

Digital interfaces offer the potential for creating new forms of accountability, efficiency, and connections between communities and operators of infrastructure. This is visible in relation to regimes of the maintenance and repair of urban systems, which are critical to the user experiences of these networks (Mattern 2018; As Idalina Bapitista (2018, 217) has argued, "the work of maintenance and repair is key to how infrastructures are continuously produced and reproduced, materially and symbolically, and sustained over time." In effect, even the most efficient, well-managed infrastructure systems are only as good as the capacity to sustain operations, and in contexts of degraded, partial, or contested infrastructures, repair and maintenance is vital, visible, and part and parcel of the everyday life of the system.

New digital interfaces between users and urban service providers in sectors such as electricity and water are creating opportunities for faster repair and maintenance in the operation of existing networks, particularly across informal spaces (Chambers and Evans 2020). Guma (2019, 3) highlighted "the crucial importance of ICT solutions in improving urban service provision through achieving operational efficiency." However, this highly depends on the urban political context and the technologies being developed. New digital tools offer the potential for the real-time monitoring of infrastructure operations to identify and respond to various problems, accelerating potential repair and maintenance and ensuring user experiences are, in the language of the digital, optimized or enhanced. The case of an app developed to improve maintenance and repair regimes of toilets in Cape Town shows that digital layering from below, developed to improve infrastructure operations, can become defunct in highly politicized techno-environments.

As I explored in chapter 5, difficulties in ensuring the safe, fully functioning operation of decentralized sanitation networks in poor, Black parts of the city has reinforced the historical and racialized ordering of Cape Town. These operational difficulties have led to moments in which the politics surrounding this vital infrastructure and the human waste it is designed to dispose of safely became central to the broader struggle for economic and racial justice in the Rainbow Nation (Robins 2016), what Colin McFarlane and I (2017) described as the "poo-litical." Out of these contested, politicized conditions, the emergence of a relatively simple digital reporting system, benefiting both the residents living in areas with serviceable toilets and a municipality that was visibly failing to maintain and repair sanitation infrastructures, was developed. The aim was to enhance the possibilities of residents in Khayelitsha accessing clean, safe, and functioning toilets. The Social Justice Coalition (SJC), a membership-based social movement mobilized across the city, and its technical support organization, Ndifuna Ukwazi, developed a campaign from 2009 called Clean and Safe Sanitation. By 2014, the SJC and Ndifuna had produced a detailed social audit of the janitorial service for communal flush toilets in Khayelitsha. These are toilets that need to be emptied by the operator every few weeks, with the waste being held on site in a storage unit.

The report by SJC and Ndifuna was damning and caused considerable discomfort to the ruling DA administration and various municipal departments. This community-led research illustrated the huge range of problems with the quality of the service being operated by the City of Cape Town, primarily in relation to maintenance and repair. This included the failure of janitors to fix or report various faults with these infrastructures, leaving residents unable to use communal toilets because of various breakdowns or restricted key-based access, forced to clean the facilities themselves, and facing conditions that were dirty and often unhygienic (Social Justice Coalition, 2014). Ndifuna and the SJC developed a simple yet potentially effective digital tool to improve these infrastructural conditions, primarily through enhanced maintenance and repair.

The City of Cape Town had already developed a reporting system for faults, disruptions, and other problems in the management of its toilets, but it was neither effective nor taken up by service users. A phone call to report a fault could take up to an hour, with the caller kept on hold by the technical operations center run by the municipality and costing around ZAR 45 (USD 3) of airtime. In 2015, the state pension of ZAR 1,350 (USD 90) per month, or the ZAR 320 (USD 21) child support grant, might have been the only income for a household in Khayelitsha. To report a fault with a toilet twice a month would have cost up to ZAR 90–100 (USD 6)—about 6–7 percent of the monthly income for a household receiving only a pension, or 14.5 percent of the monthly income for an unemployed household with two children. Evidently, the existing system did not work, with the municipality failing to provide the services, maintenance, and repair required. For many in the city, this felt like a racialized and therefore political issue rather than solely a technical problem that technical experts could fix.

Nkosikhona, from Ndifuna, made clear some problems with the current platform being used by the municipality, suggesting it was "impossible to use unless with a Telcom phone, and there are only three in the whole of Khayelitsha." And even if people reported a fault, there was no guarantee that the municipality or its contractor would repair it or ensure that it was fully serviced. A smart city this was not. In response, the two organizations set about developing a digital tool. The design of a website and reporting system using phones (smart or otherwise) would help to enhance toilet maintenance. Built through GPS mapping of the public flush toilets across the informal spaces of Khayelitsha (estimated at more than 7,000 units), the website would show the location of each toilet and its current condition. Locating the units was important, as the City of Cape Town's system was unable to pinpoint toilet locations through its own monitoring system. The rationale would be that each toilet was to have an individualized marker, which would allow residents to report a fault through a simple and cheap SMS message to the operators at Ndifuna/the SJC, who would add it to the digital interface and pass it on to ensure that the City of Cape Town attended to this report. In effect, the app had digitalized the city's toilets and made them accessible and readable as code in ways that might allow enhanced maintenance and repair.

"Louis", a dedicated worker from the municipal sanitation unit, suggested that broken toilets were being politicized and that if the SJC "report it [the fault] as it comes in, we can handle it daily, but instead they try and accumulate all the evidence against us." This suspicion across parts of the municipal administration that the toilets were being used by groups such as the SJC as politically motivated attacks against the DA was clear in the interviews we conducted. It shaped how the platform was being received and played in the racialized, political landscape and exploited so well by city leaders to justify splintered urbanism across the city.

The City of Cape Town clearly disagreed with the SJC's assessment, but on a visit that I undertook with "Louis" to PJS Section, the condition of the toilets in Khayelitsha was brought into sharp focus. It was apparent during our journey in a municipal truck known locally as a *bakkie* that they had selected the area to show me how well the sanitation infrastructure in the city in Khayelitsha was operating. I knew this because he had used his phone earlier in the day to inquire about where would be good to visit, and they assured him that PJS was ready for our visit. After only a moment, a couple of young women approached to ask what we were doing in PJS. As I explained the purpose of our visit, they let us know they were members of the SJC and one a former EPWP worker in janitorial services. The manager looked instantly uncomfortable as they described the state of the toilet block in front of us, with issues including blockages, leaking, and flooding, which had come about through the toilets being placed in the wrong location. One of our new friends was robust in her criticism of this basic mistake from the municipality, demanding, "We want the City of Cape Town to engage with the community to ask where we can put services." They explained that the last time they reported these problems through the older system, it had taken the municipality five days to repair them, creating difficulties for those living next to the toilets and for those from the surrounding dwellings seeking to use them.

As we left PJS Section, police with shotguns were walking slowly along the main road, rubbish was strewn on the nearby streets, and a few small fires were burning, reminding us of the widespread anger and resentment often expressed through so-called service delivery protests (Twala 2014). Turning the *bakkie* away from this scene to find another way out, "Louis", without prompting, reflected, "We are an abnormal department, always facing another emergency." Failing to incorporate the civic-designed digital tool for reporting faults in their operating system would surely only make their task even more difficult.

The municipality would not use this digital layer designed by the SJC to operate with the toilets of the city. This was despite its potential benefits for the hardworking maintenance and repair teams and the underfunded departments responsible for the upkeep of tens of thousands of toilet units. "Clive," a manager from the Water and Sanitation Unit, stated, "We don't have enough people monitoring toilets; if the SJC say they will partner with us, by all means we will do it." This openness to using a civil society–built digital tool and acknowledgment that the unit responsible for everyday repair and maintenance was underfunded showed how the politicians in the DA were not interested in using digital technologies to improve services if it meant acknowledging the contribution of what they

8.6   Toilets in PJS Section.

8.7   Debris strewn across the road in Khayelitsha after a service delivery protest.

considered to be an antagonistic social movement. Phumeza, the General Secretary of the SJC, lamented that the digital tool would have given "communities real-time data and supporting documentation, so it would be very difficult for the municipality to run away; they are put on the spot." The City of Cape Town failed to see such potential and, as Phumeza described, in a "pre-emptive move, they released an idea discrediting the process but had they looked at how useful it might have been, people might have responded differently."

This failure to work collaboratively with the SJC/Ndifuna was a clear indictment of how Cape Town was being governed by the DA and, in particular, the abrasive manner of the then Mayor De Lille who showed her disregard for residents in areas such as Khayelitsha on many occasions during her term. Yet, in urban worlds increasingly configured through the digital, surely resistance to incorporating these community innovations cannot continue. Nirvesh, former Chief Information Officer at the municipality, now a director of Partnership for a Digital Africa, explained how cities would need to embrace digital technologies to enhance services, including repair and maintenance. He told me, "Traditional ways of managing infrastructure have to change. How do I know what is happening out there and respond to it? We need to do it via ICT, proactively understanding and responding. That means that we started with basic administration systems. But we have to get to the next level of using e-governance to coordinate and think strategically."

In the case of the reporting app and the digital layer developed from below by activists and NGOs to deal with maintenance and repair problems across Khayelitsha's noncentralized sanitation networks, the municipality was clearly ignoring, for political motivations, a tool that could address the challenge set out by Nirvesh. This is not surprising because digital techno-environments are not being deployed in abstract space but rather are used in highly politicized conditions. In Cape Town, the deeply racialized fault lines of society and the impositions produced through racist spatial planning and infrastructural operations of racial capitalism shape the state's response to whether they adopt this digital layering. The potential of the digital to enhance regimes of repair and maintenance was therefore deeply connected to the contingent urban politics of place.

POLITICS DIGITALIZED

The rapid integration of digital technologies into the functioning and user experiences of urban networks defies narratives of African cities as playing catch-up to a Western-derived, teleological notion of infrastructural modernity. Rather, these ever-shifting techno-environments highlight trajectories beyond universalist models. Digital techno-environments are already ubiquitous but also in their infancy. Such transformations represent a disruptive opening into reconfigured infrastructural worlds amid the incessant growth of towns and cities in the region. Rapid shifts in the operation of infrastructure and the techno-environments shaped through data, code, and algorithms take place within the particular context of Africa's urbanization. Rather than ascribing the modern solely to Western cities, scholars, policy makers, and engineers, among others, need to center the experience and agency of the tens of millions of urban Africans now plugged into, navigating, and experiencing digitalized urban space.

There is growing evidence that highlights the potential for a new set of digitally generated politics and imperatives imposed from above. As the Ugandan national elections showed, this disruption can be more than technical, connecting into the political. State-sponsored control of the internet is shutting down democracy and curtailing the communicative and organizing capacity of urban populations under times of crisis in a deeply worrying and advancing trend. Thus, advocates of the democratic potential of the digital techno-environment may exaggerate in a region that has seen increasing attempts by the state to shut down these spaces. Many elections in Africa are now followed by a period of intense closure of digital realms to control urban space. These techno-environments become a site for political contestation, in terms of both the circulation of political discourse and urban political mobilizations and the ability to access the tools now required and relied upon for social reproduction.

The case of Uber in Kampala suggests new forms of digital-enabled extraction are opening up spaces of accumulation in terms of global finance, which should extend a note of caution to those celebrating rapid deployments of digital technology across urban space as a panacea to longstanding issues. As I explored, through the example of Uber, the allure of the smart city, the potential of improved urban mobility, and a functioning transport system proceed through reliance on devaluing the labor of workers, such as Joseph, the taxi driver. In effect, digital techno-environments in this form are merely the latest attempt from forces beyond the continent to profit from the poorest region on the planet and an extractive process between the center and the periphery, but also within African cities, by those who can access, accumulate, or control. Alongside surging enclaves, the capacity of the digital to enhance infrastructure and user experience is likely to benefit the few over the many.

The enhancement of existing, and yet-to-be-built, infrastructure through digital layering also highlights how the digital politics and imperatives of an urban Africa will be shaped through digital intermediation from below. Management of everyday rhythms and flows of infrastructure are increasingly refracted through software and code. And the operation of ubiquitous apps, now essential for many in navigating everyday life, is deeply connected to the urban worlds in which actors deploy them. This is despite the extractive logics of global tech corporations positioning Africa as a "blank slate" for testing, experimenting with, and piloting digital accumulation.

Rapidly urbanizing conditions, including a lack of universal, centralized infrastructures, and high levels of informality have shaped a digital techno-environment at least partly made up through "homegrown" apps, such as Mobile Money or UgaBus, and the skills, innovations, and situated knowledge of local actors such as Ronald. We must take such regionally specific stories and narratives seriously to prioritize these ways of telling and interpreting the urban future from the perspective of the phone fixer, the tech entrepreneur, or the digital activist, for such actors help to decenter the West as the sole maker of the digital urban experience, demonstrating how urban Africans are configuring their own techno-environments.

This digital layering of extant urban networks, such as transport or sanitation, offers so much potential to improve urban life. Again, we can see that urban Africa is experiencing a different pattern of technologically enabled urbanization than in previous waves, in which software and code were absent or in their infancy. Potential exists for this transformation to be mediated from within urban Africa itself as its innovators and entrepreneurs, NGOs, and social movements find new capacities, potential, and agency to interact, intervene, and reconfigure, from accelerated mobility in Kampala through the potential for enhanced regimes of repair and maintenance of sanitation systems in Cape Town to new livelihood opportunities among innovators, fixers, and hustlers, digital transformations. All precipitate new stories that need to be told about urban life. However, the example from Cape Town illustrated that those who wield power can frustrate and ultimately stop the work of actors developing this digital potential in situ and through the skills and time of those with highly localized knowledge. The promise of enhancing infrastructure through the digital does not mean it will be actualized to benefit wider publics.

The Infrastructural South is now interwoven with code, algorithm, and the functioning of digital spaces. The cases I have explored partly substantiate the promise of digital futures for the enhanced running of existing infrastructure and livelihood opportunities, even if this remains a long way from a purported techno-utopia. But they also suggest a digital layer and intensification of extraction and political and economic control as business as usual, or perhaps something more worrying.

9

POSTCOLONIAL PRESENTS
IN THE METROPOLE

FRACTURED PIPELINES AND NEW REAL-ESTATE FRONTIERS

In 2002, laboratory tests showed high levels of lead (up to 100 times the accepted Federal limits for adults, i.e., 12 parts per billion) in samples taken from local schools in Camden, New Jersey, USA. The effects of lead poisoning on children are powerful, as it is a cumulative toxin that attacks various body systems as it spreads to the brain, liver, kidney, and bones, leading to severe physical and mental development impacts. Given the well-founded public anger in Camden, the response from authorities was to cut off the pipeline supply of water. Unable to afford the substantial costs to replace the pipelines, the distribution of water switched to plastic bottles and water coolers delivered by trucks. This delivery system continued into 2019, costing the underfunded school district USD 110,000 annually.

Operations of these non-piped infrastructures are visible in Camden as the city navigates both contaminated school pipelines and the inability to provide safe drinking water to children. Roy Jones, a long-time community activist in Camden, told me, "What's amazing is the 22,000 children and 3,500 adults in every school have to drink bottled water every day." Without the possibility of a new, safe infrastructure of water provision, an off-grid system that is both incremental and peopled has operated for two decades. Amid the postindustrial decline of Camden, I saw how a part of this city, in the world's richest country, might share something in common with infrastructure in some of the popular but resource poor neighborhoods of the South than the universal, always-on and centralized systems associated with the networked city model.

Manchester in the UK was once termed "Cottonopolis," symbolizing its dominant role in the global infrastructures of cotton capitalism. The city was a place of extraordinary wealth for some. Surplus capital from these extended geographies of accumulation transformed its built environment. It was also a city in which some of the imperial infrastructures that still scaffold and inscribe the urban-regional networks of Africa were planned and lobbied for.. Indeed, "the central political issue for the British Cotton Growers Association," a Manchester-based group of cotton capitalists encouraging production in the colonies, "was public financing for railroads in the empire" (Robins 2016, 67) such as the 'Lunatic Express' line into Uganda. This global infrastructural power of Manchester was not to last. The fortunes of the city slumped from the 1930s and particularly after the Second World War as the Empire crumbled and the locus of cotton manufacturing shifted again to the global South. In the ruins of this collapse, a postindustrial city with little of its previous power or wealth stumbled forward into an uncertain future.

Perhaps the most iconic buildings of the city are the vast mill complexes in which manufacturing of cotton goods took place from supplies drawn from across the planet. The old industrial neighborhood of Ancoats was the center of this production process. Today, a real-estate partnership that includes the Abu Dhabi United Group[1] has redeveloped these mills into luxury apartment schemes. The vast petro-wealth of the Emirati elite has financed this renewal. These mills once acted as a key node in the global production of cotton. The development of these historic buildings into property assets configures them within a set of "world ecological" relations (Moore 2015) that is very different from their previous use. The formerly colonized now accumulate out of the rebuilding of the industrial city of the metropole and its rendering as a space for the accumulation of rent-seeking finance. And it's not just the Abu Dhabi ruling family that has become rentiers of Manchester's real estate. They have been joined by Chinese state and nonstate-owned construction companies, Hong Kong families and Singaporean tycoons, and Saudi private investors, among others seeking to capture extractive rents. As these financial actors from the so-called South seek new opportunities for accumulation via the secondary circuit of capital, postindustrial Manchester is emerging, like some African cities such as Accra, as a "new real estate frontier" (Gillespie 2020). It does so in ways that open up interesting parallels and divergences with the rapid proliferation of new cities and urban extensions examined already in this book. Manchester's transformation also reflects a historic shift in the financial geographies of the world economy that, like the breakdown of water infrastructure in Camden, trouble and problematize the ways in which the Global North and South have been understood as a binary across infrastructure studies.

Postindustrial cities of North America and Europe sit within the geographies of the postcolonial present as economic and political power decline and wane. The resulting *predicament techno-environments*, I suggest, force us to revisit long-held assumptions concerning infrastructural modernity, questioning whether these cities sustain the key features of the modern city through which registrars and rankings of a world of cities have been produced. These are urban spaces assumed to have modern systems of centralized, universal infrastructure configurations and to be operating as primary nodes in the global infrastructures of the world economy with accompanying sites of infrastructural power. The experiences of Camden and Manchester, and other such cities, became the basis for a model of urban modernity that was, and still is, used to assess and classify in ways that positioned African cities as unmodern. And yet amid waves of racialized austerity, utility privatization, postindustrial decline, neoliberal urban policy, and new flows of real-estate finance, global power shifts away from its anchor in the North Atlantic, and the promise of infrastructure no longer holds. SAPs are brought "home," repackaged as austerity urbanism attacking the remnants of the welfare state, public infrastructure, and the technologies of universal service provision, alongside historically prioritized relations in the global infrastructures of trade and investment.

In this chapter, I shift away from the book's focus on urban Africa to explore a relational theorization that shows how the Infrastructural South developed up to this point might be used in destabilizing existing urban theory to help rethink elsewheres, in this case what we might call the Western city. I advance calls in comparative urbanism to shift "from expected to unexpected comparisons" while "changing the flows of ideas about cities in a postcolonial urban world" (Myers 2014; see also J. Robinson 2016). I do not generate direct comparative findings from particular cities, but rather I seek to open up a relationally informed analysis that is open to the possibilities of mobilizing the ideas and vocabularies generated out of previous chapters focused on the Infrastructural South. The aim is to suggest that assumptions about urban infrastructure in the Global North also require thinking anew. Ideas that cities such as Manchester and Camden have reached an end point of networked city and logistical power status no longer hold. In doing so and returning to where I first began to think about cities, I challenge assumptions about these systems that stretch out not just to urban Africa but also into the "modern" cities of the West.

In this mode of relational, comparative analysis, the Infrastructural South again acts as an epistemological space in which to decenter Western cities. As Jochen Monstadt and Sophie Schramm (2017, 123) helpfully suggested, "The study of infrastructural landscapes in urban Africa provides insights for better understanding recent changes in the technological fabric of cities of the global North." This comparative practice has already contributed, complicated, and extended understandings of infrastructure (e.g., Ranganathan and Balazs 2015) as a way in which we can begin "thinking cities through elsewhere" (J. Robinson 2016).

REVISITING URBAN MODERNITY IN THE WEST

Infrastructurally, I contend that the postcolonial present is characterized by disinvestment, breakdown, and precarious, disposable lives, alongside the waning power of logistical and trade networks and uncertain futures. Collectively, these conditions destabilize a priori assumptions concerning the global hierarchy of cities. One starting point in outlining this present is the overlapping ways in which SAPs in African cities and austerity urbanism in Europe and America have proceeded in recent decades. As Reuss (2015, n.d.) contended, "A rose by any other name would smell as . . . catastrophic." Infrastructure becomes a productive site for thinking comparatively across seemingly diverse urban contexts that blur and complicate conceptions of North and South. Concurrently, we can mobilize ideas generated through engagement with the South to think about a revised infrastructural geography of the Western city. This is a reversal of knowledge production practices that should be and indeed is being embraced in urban studies (Myers 2020).

Amplified through the 2008–2009 financial crisis, the experience of SAPs in urban Africa has found expression across the cities in Europe and North America through waves of austerity that have intensified longer periods of neoliberal restructuring, racial capitalism, and postindustrial decline (Davies and Blanco 2017). We can see this in the lack of investment and subsequent collapse of safe, fully operating public infrastructures in US cities such as Flint (Pulido 2016 Ranganathan 2016) and Detroit (Ponder and Omstedt 2022). Given these conditions, some commentators, perhaps problematically, question whether these spaces can even be considered Western, reflecting particular (and racialized) ideas about how urban conditions have come to be understood. For instance, Cowie (2001, 205) commented on Camden, "Today the south Jersey city may be more akin to the Third World."

It is not just disinvestment in public infrastructure that suggest a convergence of underlying operations and geographies and the blurring of conceptual distinctions between North and South. The global search for investment by municipal and business leaders to finance the rebuilding of postindustrial cities increasingly involves securing new flows of capital from the South. Finance from those regions of the world that had once been subjected to imperial control, extraction and exploitation are now underwriting the rapid transformation of urban spaces in the metropole. As real-estate markets become globalized, financial power finds new urban spaces from which to generate rents. This has proceeded primarily through securitization, which "is a major financial invention that allows real estate to be financed in global securities markets that are disconnected from local property markets" (Gotham 2006, 235). Increasingly, cities such as Manchester are being rebuilt through finance from beyond the West, questioning the position of these cities in traditional hierarchies and rankings of global economic power. These new financial geographies of the urban and infrastructural indicate the predicaments facing policy makers and politicians as they attempt to navigate uncertain and turbulent futures.

Let's follow a mode of thinking that structural adjustment and the austerity dogma it engenders have come home, back to the North Atlantic heartlands of imperialism and neoliberal doctrine. History is repeating itself (with a difference), despite the seemingly obvious lessons learnt across

Africa and beyond of the ruinous effects of public disinvestment. As Ugandan activist Kalundi Serumaga (2015) argued in relation to Greece, "If the country that birthed the European civilization can no longer afford to live by European standards, then who can? Not Africa, surely." Structural adjustment has taken on a different guise in the metropole, primarily through what is understood as austerity, with clear urban and infrastructural dimensions (Bayırbağ, Davies, and Muench 2017) that some have termed "austerity urbanism" (Peck 2012; Tonkiss 2013). They also share much in common (despite many divergences in histories, conditions, and politics) with the SAPs through the irrational, technocratic obsession with economic growth, liberalization, and cuts to social spending, including in urban service provision. Infrastructural futures then may have been prefigured in seemingly peripheral regions, including urban Africa.[2] Such a proposition challenges notion of a completed Western urban modernity as infrastructures come under various forms of pressure, breakdown, and reconfiguration.

Collectively, these structural changes to urban economies take on the form of what Naomi Klein (2007) termed the "shock doctrine" imposed in a productive moment of financial crisis (in this case, the economic shocks of 2008/2009 or the longer postindustrial decline in the period following the Second World War) that pushed debt to supposedly unsustainable levels. The effects have tended to be catastrophic. Austerity programs have led to a reduction in the effectiveness of welfare services such as health, education, urban service provision, and the capacity of the state to invest in and shape the built environment. In what follows, I think through these postcolonial presents from an infrastructural perspective. I do so to show how, given a similar mode of thinking that has helped outline the idea of the Infrastructural South, that the infrastructure of the so-called Global North does not match the characterizations often ascribed to these spaces.

THE END OF THE NETWORKED CITY?

PIPELINE CRISIS

Camden, a city of 80,000, sits across the Delaware River from its bigger neighbor, Philadelphia. This New Jersey city has pronounced infrastructural challenges above and beyond the contaminated school pipelines. Racialized austerity, underinvestment, and subsequent decay have generated everyday injustices for communities across the piped infrastructures of water and sewerage in Camden. Across the US Northeast and Rust Belt, as cities face an intensifying pipeline crisis, these conditions in which universal, safe, and fully functioning infrastructure may no longer be certain become visible. Flint, Michigan, has perhaps been the most high-profile example of this pipeline crisis and subsequent impacts on poor, often Black and Brown populations (Pulido 2016). But as a Reuters investigation found, more than 3,000 municipalities across the USA have water pipelines with high lead levels (Pell and Schneyer 2016). And it's not just lead contamination that is causing a pipeline crisis. From the water shutoffs in Detroit (affecting 160,000 households between 2014 and 2021) to the reemergence of waterborne diseases such as Legionella, with more than 20,000 cases recorded between 2000 and 2009 (Centers for Disease Control and Prevention 2011) to the 14 million American households unable to afford water (Mack and Wrase 2017), the capacity to provide fully functioning water supply and safe sanitation services is under severe stress.

Underlying this pipeline crisis is the public expenditure required to address long-term decay across municipal water, sanitation, and storm-water systems, estimated at up to USD 1 trillion just on necessary infrastructure for drinking water (American Water Works Association 2012). Even with promised Federal funding from the Biden government of up to USD 55 billion, fixing these pipelines will require significant utility bill increases, leaving one in three households struggling to pay for water (Mack and Wrase 2017). Across the ageing infrastructures and toxic environments of these postindustrial cities, the long-held public good of universal service provision, the promise of infrastructure, and the certainties of the networked city model are, like the rusty pipelines themselves, disintegrating.

FROM "CITY INVINCIBLE" TO "THE LAST STOP ON THE WAY TO HELL"

Camden is the last stop on the way to hell . . . It is the last stop for the shit too.
—Raymond, Latin American Economic Development Association

"In a Dream, I Saw a City Invincible" wrote Walt Whitman (1867) about Camden, an archetypal industrializing city of the nineteenth century. Public infrastructure emerged in industrial Camden from the requirements of capital to sustain the social reproduction of labor and to address the demands of the working class. It operated to control and bind the techno-environments of the "bacteriological city" (Melosi 2000). The Camden Water Works Company was incorporated in 1845. In 1846, municipal actors drew plans up for the first supply system. These included a wharf along the River Delaware, a brick building on Cooper Street, piping, a 10-horsepower steam engine, and attendant pumps (Public Ledger 1846, 3). A private piped system in 1853 allowed for water to "be conveyed into the city by a capacious and magnificent aqueduct, a distance of about two and a half miles" (Public Ledger 1853, 1).

In 1870, the City of Camden purchased the water system for USD 200,000 as the State granted the municipality "exclusive right of furnishing water to the citizens of Camden" (State of New Jersey 1871, 415). Municipalization ensured public ownership and control over infrastructure as the city expanded. In 1886, the city increased the size of the pumping and distributing capacity, and more than 5 million gallons a day flowed across forty-six miles of pipeline, generating more than USD 75,000 annually (Prowell 1886, 439) and highlighting how the municipality was willing to invest in the public infrastructure of the "modern city" (Graham and Marvin 2001; Tarr 1984). Alongside water, the municipality established a sanitation network in the 1880s through the creation of a combined sewerage outflow (CSO) system. It was designed to treat sewage flows, as well as contain rainwater runoff and wastewater. These publicly owned water and sanitation networks remained relatively unchanged during the twentieth century. They offered universal, safe, and fully functioning services, operated by municipal departments in the city itself and surrounding suburbs and financed by the growing population and ratepayer base. This infrastructural landscape, typical of the model of the networked city, was to change profoundly from the 1970s onward.

Camden experienced many of the hallmarks of decline familiar across the post-Fordist landscapes of US cities (Marcuse 1997). Howard Gillette (2005) described this urban history simply as "the fall." The economic collapse of the city led to pronounced racial segregation between Camden and surrounding suburban Camden County like in many US metropolitan regions. White flight accelerated as deindustrialization bit, jobs disappeared, and disturbances occurred in 1969 and 1971. African American

9.1 Camden has many of the hallmarks of a postindustrial city in decline.

and Puerto Rican communities were established, changing the population significantly. This demographic transition reached a peak between 1970 and 1980 during which Camden shifted from a majority white city (falling from 60 percent of the total population to 31 percent) to a majority Black city (rising from 39 to 53 percent).

Raymond, a business leader, explained that the segregation between city and suburb was now "racially and economically very stark," with the white population in 2010 comprising only 17 percent in the city compared with 65 percent in the surrounding suburbs (US Census Bureau, n.d.). The collapse of industry and losing so many taxpayers hit Camden hard, leading to nearly 40 percent of residents living in poverty by 2010 (US Census Bureau, n.d.). Popular representations of the city frame it as a paradigmatic example of the collapse of the Western industrial city, which were only amplified by statistics such as "America's highest per capita murder rate in 2012" (Mathis 2015).

It was this profound decline that led Raymond to describe the city as the "last stop on the way to hell." The shift to a Black and Brown city surrounded by a jurisdictionally and fiscally separate, white, suburban hinterland meant that by the late 1990s, Camden was at risk of bankruptcy (declared by Mayor Milan in 1999) as its ratepayer base fled out of the boundaries of the municipality and a regime of racialized austerity was implemented. In 2001, state-aid contributions accounting for 65 percent of the total operating budget sustained the city. These austerity urbanism conditions of the 1990s in which the city sold off many physical assets (Gillette 2005) would have far-reaching effects on the infrastructure geographies of the city, including setting the conditions for the privatization of the water and sewerage system in 1999[3] alongside systemic disinvestment and disrepair.

IS CAMDEN A NETWORKED CITY?

What are these conditions that might suggest Camden is no longer a fully networked city? First, residents in various neighborhoods of the city, and particularly Waterfront South, experience outflows from the CSO when the system reaches capacity. Wastewater and sewage are pushed above the surface, incorporating "high levels of suspended solids, pathogenic micro-organisms, toxic pollutants, floatables, nutrients, oxygen-demanding compounds, oil and grease, and other pollutants" (EPA 1994). "Peter," a Camden County Municipal Utilities Authority (CCMUA) official, I spoke to noted that the "burden of cleaning the rivers was put on 1,800 people" in the Waterfront South neighborhood because the County Regional Wastewater Treatment System was located nearby. The pipelines carry sewerage from the thirty-seven municipalities of Camden County of up to 58 million gallons of sewage a day, which is prone to spilling out onto streets, given it "essentially has no line capacity for additional sewage during wet weather periods" (New Jersey Futures 2014, 86). This was not helped by a system that was, according to "Peter," "dilapidated and poorly maintained." He described how the community remains a "poster child for environmental injustice in the state," not only for the treatment plant but also for the nearby county-waste incinerator, which burns 1,000 tons of trash daily. And these outflows have also caused problems for other neighborhoods in the city even as the CCMUA has led attempts to manage this challenge within constained budgets. Raymond explained that everyday life in many parts of Camden meant experiencing the effects of "the poop that everyone

9.2   Results of a CSO outflow in a Camden neighborhood.

else is pumping from outside the city." More than seventy overflows were recorded in 2017, violating the Federal limit of four (EPA 2018).

Second, severe stress has been faced by the city across the pipelines of the water system because of a lack of municipal investment and capacity to finance repairs and maintenance. Obsolete pipelines were common; some were even made of wood and were more than a century old, generating a series of technical issues in providing a safe supply. As Hall, Lobina, and Corral (2010, 4) wrote, "Camden residents have long complained of poor water quality and brown water from their faucets . . . the potential for contamination in the water system is grave." This physical decay of the system—including the rusting of pipes—compounded by a lack of maintenance and investment, has made the water supply a potential health hazard for households. The lead contamination present in school pipelines was merely one expression of these problems and led to alternate arrangements both in and out of the schools. Here, water supply has become noticeably off-grid and informal, sharing more in common with the incremental urbanisms of the Infrastructural South than with the networked city model. Baptista (2019, 515) argued about electricity in Maputo that "the practice of delivering the service itself requires considerable creativity in the translation of the ideal into situated, context-specific alternative arrangements." In Camden, these translations are also present through the nonnetwork system that became and remains operational, enabling supply beyond the main network and peopled by "water carriers" previously understood to be present only in cities of the South. Kooy (2014, 34) argued that "informal water providers are, thus, assumed to gradually disappear through the growth of the urban infrastructural ideal." Without the financial capacity to repair existing pipelines, Camden's non-piped and peopled water supply asks us to question whether the modernist vision of infrastructure might have partly collapsed.

Third, Camden's water supply has regularly been interrupted. This is partly due to low pipeline pressure because of the neglect over previous decades. In one incident in 2016, the utility issued a boil-water advisory to more than 40,000 residents, roughly half the city's population, for activities such as cleaning teeth. A fault with a crucial pipe led to neighborhoods west of the Cooper River losing pressure. Camden continues to experience interruptions regularly, with little evidence to suggest that the same service problems occur in surrounding predominantly white suburbs. For example, in neighboring, overwhelmingly white Collingswood, service users can access drinking water that "meets or exceeds all federal and state monitoring requirements" (Collingswood Water Department 2017). The municipal water department has created a capital investment plan to "upgrade our existing treatment plants, replace undersized water mains and water-service connections from the street to the curb." As I have set out in thinking about the Infrastructural South as condition, disruption and breakdown have long characterized the lived experience of urban majorities and have been juxtaposed against the seeming constant supplies of cities in the Global North. Indeed, Truelove's (2011, 147) description of Delhi's water supply—"categorized by the intermittent hours that water runs, insufficient and irregular pressure of water when it is running, sudden breakdowns . . . and problems with contamination"—could easily be describing the conditions in Camden.

Fourth, lack of maintenance and investment in the water system meant high leakage rates of up to 45 percent reported during the 2004–2008 period, even though this was supposed to be limited to 10 percent (NJOSC 2009). Leaking from ageing pipes was a critical factor. As a utility worker said, "There were sections [of pipelines] that were not jetted or cleaned for ten years." The New Jersey Office of the State Comptroller described conditions as "more comparable to that of cities in developing countries" (NJOSC, 2009). An extraordinary lack of private contractor knowledge complicated approaches to addressing leakage, notably the failure to know the location of parts of the actual infrastructure. The NJOSC (2009, 1) reported that officials "attempted to locate seventeen city-owned assets purportedly maintained by United Water—including pumps, valves and hydrants—and could not locate fifteen of them." In Mumbai, Anand (2017, 24) observed, "The prolific leakages of water from the city's underground network" showed how infrastructure is a "living, breathing, leaking assemblage of more-than-human relations" (6). Understanding leakage in this way reframes the losses from Camden's water system as a predicament shared across not just the South but also the crumbling postindustrial cities of the North. Leakage is therefore another way in which we might understand Camden as no longer a networked city, challenging assumptions about the global geographies of infrastructure.

Fifth, compounding the already substantial problems faced by residents in accessing the water supply in early 2019, the privatized utility proposed a water shutoff program. Initial implementation would target roughly 400 households in arrears. Such a program meant Camden would join cities such as Detroit, where thousands of residents can no longer access water because of poverty, generating concerns about a potential future health emergency. As Food and Water Watch (2010, n.d.) argued, "These shutoffs threaten public health, community wellbeing and basic human dignity . . . Without running water, people cannot cook, clean, shower, wash their hands or flush their toilets." Such shutoffs would echo attempts by companies such as Suez to force through cost-recovery schemes as part of a two-decade wave of accumulation across African cities, such as Johannesburg (Narsiah 2011). Pauw (2003, 819) described the effects of water shutoffs on residents in South Africa, "forcing them to get their water from polluted rivers and lakes and leading to South Africa's worst cholera outbreak." If Camden has yet to suffer such a public-health emergency, the potential remains if American Water pushes forward with disconnection for a similar level of health crisis.

Camden is a glaring but not untypical example of the various pipeline problems facing postindustrial cities in the US Northeast and Rust Belt.

This predicament techno-environment suggests we can no longer characterize the infrastructural geography of Camden through the networked city model and need to think anew about our theoretical assumptions. However, it is not just the breakdown of service provisions that are challenging our ideas about the urban infrastructures of the North.

RE-WORLDING THE POSTINDUSTRIAL CITY

New circulations of rent-seeking finance in postindustrial cities offer opportunities to generate and actualize futures in urban spaces that had previously experienced decline and public disinvestment. We can understand these transformations as a form of "worlding" (Roy and Ong 2011) or, given the past role of many of these cities in the global economy, re-worlding. I use this term to highlight how these left-behind cities are being made global again through surging flows of international real estate and private infrastructure finance from the East or South and which differ dramatically from what has gone before in the metropole.

A visit to China by then UK finance minister George Osborne in 2015 symbolized an important moment in the recent history of attempts at "re-worlding" cities in regions such as the north of England. He said, "This week I have come here with my friends and colleagues from many different cities in the north of England to promote investment in the Northern Powerhouse." In this attempt to secure investment, Osborne's desire to use Chinese finance to rebuild cities rather than public funds highlighted both the inability of the UK state to undertake the task and how the geographies of the world economy had irreversibly shifted away from the North Atlantic. Osborne was primarily in China to offer new opportunities for Chinese investors, particularly through real-estate developments in the deindustrialized north of England. The visit culminated in Osborne, on the last day in Chengdu, brazenly issuing a catalogue for potential Chinese investors of various opportunities. After years of massive austerity programs, curtailing the capacity of the state to rebuild and regenerate a new phase of neoliberal intervention was necessary (Davies and Blanco 2017). Crucial to this strategy was tappin into the ever-increasing flows of global real-estate finance. Later in the year, China's president Xi Jinping joined Osborne in Manchester, signaling a new phase

9.3 Crowds gather outside Manchester Town Hall for the visit of Xi Jinping.

in historic relations between the countries in which the UK was now the junior partner.

Rapid transformation in the urban geographies of cities such as Manchester, which traditionally held so much global infrastructural power, demonstrates the postcolonial presents that now characterize cities of the North. This, I contend, is another way in which predicament techno-environments are being materialized and another example of how present conditions are destabilizing binaries of what the infrastructures of the city mean in a global geography. Walking through Manchester over the last few decades has been an encounter with the ruins and reminders of imperial infrastructures at the urban scale. Various new financial actors are now transforming these sites again, instigating a new era of economic relations between the city and the world. As these shifts proceed, we need to question again how and where we locate the Infrastructural South.

The metropolitan region of Greater Manchester, England, has long been associated with the rise and fall of industrial capitalism (Engels 1892; Hodos 2011). The history of the conurbation shows how formerly powerful urban spaces in the metropole experienced sustained decline. In recent years, some of these postindustrial cities have emerged out of this decline through built environment transformations that are often financed via newly assertive economic powers in the South. This finance, often derived from natural resource accumulations such as oil, seeks to invest surplus capital into real estate in the metropole. These new, extended processes of urbanization highlight the transforming historical relations between (former) metropolitan centers and peripheries, reversing flows of finance and power that suggest a profound reconfiguration of the postindustrial city. Like the private equity company Rendeavour operating in Ghana and using finance generated through extractive activities in Siberian forests to build Appolonia, cities such as Manchester are increasingly being rebuilt through these new circulations of rent-seeking capital derived from the South.

In ongoing work that I have undertaken, most recently with Rich Goulding and Adam Leaver (2023) on the real-estate boom in central, high-density areas of Greater Manchester, we found significant flows of international investment in the city's build environment. Over the period 2012–2020, of the 45,000 apartment units built, 13,609 (26 percent) involved international finance, with nearly 50 percent of this originating from Hong Kong, China, United Arab Emirates, and Saudi Arabia. And this was significantly higher in the rental sector, in which international finance is more easily integrated into the urban development process. This investment represents a relatively recent phenomenon in the city, brought about through its incorporation into global flows of real-estate investment and the policies of the region to attract international finance to support its postindustrial rebuilding.

THE SILK ROAD TO SALFORD

Middlewood Locks is in the City of Salford (one of ten boroughs that make up the Greater Manchester metropolitan region). Since the late eighteenth century, the site had operated as a space of global infrastructure, orientated toward and actively used throughout the industrial era before a decades-long decline and subsequent attempts at establishing new economic purpose. In recent years, new actors in the city have transformed this infrastructural wasteland into an enclave in which hundreds of new apartments are being built through Chinese (and Singaporean) investment, connected to the BRI, like the corridor investments explored in chapter 7, and that have effectively brought the Silk Road to Salford.

Middlewood began its urban history from agricultural land after the American Revolutionary War in 1793 as investors saw renewed opportunities to use infrastructures for the cotton trade, heavily interrupted by North Atlantic hostilities. The establishment of the Manchester, Bolton, and Bury Canal began in 1795, with investment of £47,000 from nearly 100 investors. This part of the city was shaping into a globalized production space as capitalists put new technologies to work to enable the circulation of raw materials and produced commodities. As Maw et al. (2012, 1495) argued, "Five new public canals and 23 private canal branches activated a major expansion of Manchester's waterfront, providing the majority of the manufacturing sites that enabled the town to become the world's foremost factory center."

In the 1820s, William Higgins and Sons constructed a cotton mill to take advantage of these globally facing infrastructures that extended and connected to networks of extraction, distribution, and mobility. The city was becoming a critical site for global production networks that were being actively shaped through imperial rule over new commodity frontiers. The business later specialized in the development of new technologies for the cotton trade, with these technical innovations establishing new forms of standardization, particularly around the manufacture of new spinning machines.

The opening of George Stephenson's Liverpool and Manchester Railway in 1830 from the nearby Liverpool Road Station would herald a revolution in transport infrastructure that would quickly come to supersede the speed, scope, and capacities of the canals. The railway line that would go on to be acknowledged as the world's first intercity passenger service traveled along the southeastern periphery of the Middlewood Locks site before the construction of the Ordsall Lane station in 1849 provided a direct connection for local industry and the rapidly growing local population. In 1848, the site had become increasingly aligned with the opportunities present in the new railway age with the establishment of the Salford Rolling Yards constructed along Hampson Street. P. R. Jackson and Co., established in 1840, focused on machinery that could produce railway tires and was celebrated by organizers of the 1851 Great Exhibition. Manufacturing, technical innovation, and new global infrastructure located in Middlewood and the wider metropolitan region predicated an active demand to expand the geographies of trade and circulation through what Jerome Hodos (2011) termed "municipal foreign policy." This global infrastructural power was underpinned by what Sven Beckert (2015, 86) described as "a powerful and influential pressure group in their efforts to acquaint the planters and the British Government with their requirements." As Keller Easterling (2014, 99) argued more broadly in relation to global infrastructure, "the capacities of these new technologies together with the territorial conquests of the private infrastructure lent extra powers to, and even inspired, the ambitions and policies of these states."

Middlewood Locks provides one of countless sites that shaped urban and infrastructural space in industrial cities in the North Atlantic and relationally across those regions termed the South. Built through new technologies that enabled the appropriation, extraction, and exploitation of global commodities, including canals, industrial manufacturing and processing, and railway technologies, these were sites of global infrastructural power. Middlewood Locks, like the British Empire, spun into decline and later ruin. The city fell to the margins of global networks of trade. Authorities demolished Ordsall Lane station in 1957 as the British Government reduced the capacity of the rail network. Dwindling supplies in the Lancashire coalfields meant the canal was abandoned in 1961, falling into disrepair and eventually being filled in and covered. Salford Goods Yard would not last much longer, closing in the 1970s. Housing for the working class was demolished and populations displaced. The Rolling Mills, run by P. R. Jackson and Co., involved in much of the innovation and development of railway infrastructure for more than 100 years, carried on into the 1980s before closure and demolition in the early 1990s. This urban space, once an engine of industrial capitalism and a site of global infrastructure power, had reached a nadir. Like the wider city, it experienced a relegation to the periphery of the world economy as those "cheaper" regions of the world took on various new economic functions, roles, and eventually power.

Middlewood Locks remained a large, former industrial zone facing across the River Irwell toward central Manchester for many years, surrounded by similar areas of former industrial use that sat idle and unused. After various attempts at transforming the site, including a failed snowdome, in 2013, the UK-based Scarborough Group produced new plans and sources of financing for Middlewood Locks to shift expectations over its future. This masterplan sought to develop the site into a neighborhood with canal-side apartments and commercial space, including offices and hotels, and made attractive to investors through easy access to central Manchester across the River Irwell, following a global model for large-scale urban real-estate transformations. The development was subsidized by the local state through public investment in site remediation (around USD 6 million) and in waiving financial contributions from the developer, required as part of the planning process (around USD 9 million). This subsidy reflects how austerity urbanism has proceeded in the UK as the functions of the local state shift toward creating enabling conditions for real-estate investors, often at public cost.

Developers have extensively rebuilt the site, with many of the 2,000 planned housing units already constructed and proposals for 750,000 square feet of commercial space, a hotel, and various leisure uses likely in coming years. The urban development has reflected the wider metropolitan real-estate market, which has become increasingly financialized and internationalized. Developers envisaged the site as space for high-end "urban living" like Appolonia or Pearl in the City in Accra. The site transformed into a whole new urban extension constructed at a dizzying speed. What might have surprised anyone visiting during construction were the Mandarin language hoardings and flags across the site that highlighted the role of various Chinese actors in the remaking of this former site of global infrastructural power.

9.4 A Beijing Construction and Engineering Group (BCEG) hoarding at Middlewood Locks.

9.5 Cranes stand over the Middlewood Locks development.

NEW ACTORS

In order to secure financing, the Scarborough Group established Fairbriar International as a joint venture vehicle. It comprised Scarborough International Properties with a 50 percent holding, alongside Metro Holdings of Singapore and Hualing Industry and Trade Group of China (hereby Hualing Group) both of whom control 25 percent. Fairbriar estimated the total gross development value at around USD 1.5 billion, meaning tens if not hundreds of millions of dollars being invested by the international partners. Fairbriar awarded Beijing Construction and Engineering Group (BCEG) the construction contract. BCEG has become a highly visible actor in infrastructure in the region through its involvement in the Airport City zone and construction on the Hong Kong–based Far East Consortium's massive Victoria North real-estate project.

It is the investment from the Hualing Group and the awarding of the construction contract to BCEG that perhaps best convey a connective pattern of urban development across the real-estate frontiers across South/North in which the role of Chinese finance has become integral to contemporary urban transformation whether in Middlewood or in Fortune City, Accra. The Hualing Group is a private-sector business based in Xinjiang province that began in 1988 developing various markets, including wholesale in Ürümqi. This business soon formed a construction and development company to build these markets and has subsequently received active support from the Chinese state to develop new opportunities overseas. The recent surge in infrastructural projects in Xinjiang led to companies such as Hualing Group being supported by the Chinese state to expand internationally. They have been investing profits created over the last decade or so of state-led economic growth into new markets. Previous investment by the Hualing Group has focused on assembling segments of massive global infrastructure within Xinjiang and then beyond China's borders. This has included a free trade zone in Kutaisi City designed to connect the capital of Georgia, Tbilisi, with the Black Sea Ports of Poti and Batumi. There is a range of other infrastructure corridor projects established by Hualing Group in Georgia, including Tbilisi Sea New City and a series of new hotels. Involvement in this geo-strategically important country in the Caucasus region of Eurasia and at the crossroads of Western Asia and Eastern Europe dates back to 2007. Investment by the company now totals more than USD

500 million. The announcement of investment in the UK from Hualing Group, including in Middlewood, was closely aligned to the opportunities of the BRI, drawing Salford and the wider Greater Manchester region into the New Silk Road. It has come as part of what a public investment official estimated as up to USD 2 billion of Chinese-based investment into real estate in the region in the last few years.

The second key actor involved in Middlewood is a Chinese state-owned enterprise that operates as a developer and construction company. Established in the 1950s delivering Chinese foreign aid projects, BCEG has grown to become a top 50 global construction company that is active in more than twenty countries, especially in sub-Saharan Africa, including Congo, Angola, and Rwanda, and through projects such as the Tanzania National Stadium. BCEG established its initial presence and UK headquarters at Airport City, Manchester Airport. This was a new development incorporating offices, hotels, and other functions and billed as a new global business district. As a BCEG official explained to research collaborator Alan Wiig in our joint research project, "China is a new player in the global market in construction" and seeking new opportunities. The company was attracted to the UK because the government "wanted to open up to more investments from China." BCEG's involvement in Middlewood Locks and elsewhere shows how they have become entangled in both infrastructural and real-estate projects in the metropolitan region. As the former Chief Executive of Manchester City Council said on the firm's website, "they are now widely acknowledged as a major player in the region."

The historic metropole and its redundant sites of global infrastructure power are now being "re-worlded" through finance from the South, demonstrated by the role of the Hauling Group and BCEG in Middlewood Locks and countless other examples across the metropolitan region, such as the involvement of Abu Dhabi in the repurposing of former cotton mills. As the UK Finance Minister made clear when announcing the Middlewood Locks investment in 2015, "We are building an ever-closer relationship with China"—a sentiment shared by "James," a public official involved in investment in the city, who stated to Alan that, "China is the top priority in our internationalization activity."

Given these circuits of real-estate finance, expertise, urban development models, construction companies, and associated international relations,

Manchester has, like Accra, become a "new real estate frontier" (Gillespie 2020). The city has become integrated into international investment flows, and the local state has used former sites of global infrastructure power and public funds to attract this investment. Materially and financially, the historic primacy of Western cities in the operations of the world economy has been turned upside down. These spaces formerly acted as key nodes in imperial geographies, the capitalists that owned the industries on these sites coordinated and accumulated from extractions from the South. Now, real-estate investment has transformed these same sites into spaces of extraction for the so-called periphery. This is a profound reversal in fortunes that has decentered and destabilized how we think and understand the North/South binary across urban and infrastructure studies.

BEYOND AN INFRASTRUCTURAL BINARY

In this chapter, I set out to show how the need to think anew about the role of infrastructures in Africa's urbanization can also be extended. Here, I do not mean to claim Camden or Manchester as part of the Infrastructural South, but rather to show that the need to develop new explanations, analysis, and conceptual understandings of infrastructure is not confined to African cities but rather is global, as old assumptions and binaries about urban modernity crumble alongside pipelines and logistical power. I have shown that postindustrial cities in the North no longer reflect an infrastructural modernity of universal fully functioning services, as well as global infrastructural power denoting primary, coordinating roles within the world economy. The results trouble existing ways of theorizing urban infrastructure and cities more broadly. They have shown how the imperatives to articulate the Infrastructural South as a condition and epistemology are also at play in different urban contexts.

First, I considered whether the networked city is sustained amid widespread decline and disinvestment. Since introducing publicly run, networked services in Western cities, the model of the networked city has become central to the idea of urban modernity (Dupuy and Tarr 1988; Graham and Marvin 2001; Kaika and Swyngedouw 2000). We can no longer take for granted assumptions about the infrastructure of the Western city as universal, safe, and fully functioning, given the findings from

Camden and similar conditions in other US cities. This was a predicament techno-environment in which municipal officials, racialized communities, and utility operators faced issues such as high leakage rates, improvised, nonnetwork infrastructure provision, unsafe drinking water, toxic CSO outflows, and ongoing disruption in supply. In Camden, the capacity to operate a safe, fully functioning infrastructure is experiencing severe pressure, if not outright collapse, because of the effects of long-term, racialized processes of disinvestment, disrepair, and breakdown throughout the US Northeast and Rust Belt. Despite assorted state and non-state efforts to reverse decline, this is a growing trend that suggests the actually existing networked city may not return soon.

Global urbanization follows multiple and fragmented trajectories in which parts of the South now far exceed the infrastructure operating in cities such as Camden. Think about the always-on, guaranteed infrastructural services of Appolonia or Pearl in the City. Concurrently, other urban spaces in cities such as Accra or Kampala, with partial provision and disrupted infrastructure, share interesting similarities with Camden (despite obvious differences in histories, governance, and techno-environments). Struggles for survival across infrastructure and within resource-poor communities are not just located across cities in the South. The challenge for cities to operate basic infrastructure services is not confined to the African city. Camden illustrates how the struggle for basic rights through fully functioning, safe infrastructure crosses the geographical and theoretical binary of North/South. Basic human rights of communities are not being fulfilled in the shattered infrastructural worlds of Camden and other prominent cases such as Flint and Detroit. Through Resolution 64/292, the United Nations General Assembly (2010) recognized the human right to water and sanitation, acknowledging clean, safe drinking water and sanitation underpin the realization of all human rights. If the justification for extensive involvement of agencies such as UN-Habitat in urban Africa has been to provide extensive financial and technical support to governments and municipalities, might it be time to call for such United Nations assistance for cities in the USA?

Second, I questioned whether the global infrastructure power of the former metropole was still clear and operational across the built environment of the postindustrial city. Again, as in thinking through whether the networked city model has been sustained, the results challenged how we come to describe and analyse infrastructure. In Greater Manchester, in spaces such as Middlewood Locks, Salford, the built environment is being rebuilt and repurposed. This transformation shows how global infrastructure power is no longer centered in the postindustrial city. Various logistical networks, spaces of industrial capitalism, technological innovation, and associated functions have long disappeared—and the coordinating role of the city in the global economy long departed. This site has now become a space of speculative real-estate development as the city is "re-worlded" through incorporation into global circulations of rent-seeking finance. Urban development in Manchester, despite major divergences, shares commonalities with those in Africa, such as Accra, which "are seeking to replicate Asian 'worlding' [ibid.] strategies with ambitious designs for large-scale master-planned urban projects" (Gillespie 2020, 603).

Given the austerity urbanism of the last decade, municipalities and national governments in postindustrial contexts have increasingly encouraged and relied upon new flows of finance to fund transformations into the built environment with the hope of wider investment in the city. The emergence and involvement of new global actors from countries such as China to Abu Dhabi in this financing symbolize a definitive shift in how and who remakes the built environments of the metropole. Out of these geographies, we can understand Manchester as a site of extraction by rent-seeking finance from the South, rather than a space of global infrastructure power coordinating and profiting from such extraction elsewhere. The former metropole has to an extent become the periphery.

The breakdown of the networked city and the surging flows of real-estate finance rebuilding disused sites of global infrastructure power challenge all of us to think anew about the assumptions of infrastructural modernity. Contemporary conditions in some postindustrial cities make redundant binary understandings of modern/unmodern and North/South. The logics and rationalities of predicament techno-environments being produced in urban contexts, such as Camden and Manchester, require rejecting a completeness and techno-environmental end of history. It is not only in urban Africa that it is "better to see infrastructures as emergent, shifting and thus incomplete" (Guma 2020, 728). This chapter suggests the Infrastructural South as a concept might open up new conversations about how we think about infrastructure elsewhere, including the "modern" Western city.

10

TOWARD A POPULAR INFRASTRUCTURE

NAVIGATIONS

Vodunaut is a series of pieces produced by Beninese artist Emo de Medeiros: a set of traditional textured seashell headsets refashioned into space helmets displaying video works via smartphones. De Medeiros blends together an interest in cowry shells, symbolizing voyage, with space navigation and charting the future through the Fa philosophy and geomancy practiced in Benin. The artist (De Medeiros, n.d.) emphasized, "Its particularity [which] resides in the fact that it not only describes an array of possible futures but also how one should navigate them." Bringing together these histories of craft, philosophy, and exchange with the digital and outer space makes visible "transcultural spaces and the questioning of traditional notions of origin, locus or identity and their mutations through non-linear narratives" (Levontin 2020). This is a material assemblage that is bound outside the logics, rationalities, and trajectories of Eurocentrism and built as an engagement in speculative futures and African ways of knowing.

Achille Mbembe's writing echoes de Medeiros's fascination with navigations of journeys, routes, and explorations. He develops these ideas in part by engaging with the long-held traditions and movements of the Dogon people (of present-day Mali), which Mbembe (2018, 4) suggested "could lead to diversions, conversations and intersections" that are "more important than points, lines and surfaces which are as we know cardinal references in western geometrics." The Dogon's origin stories of interstellar travel[1] and astronomical knowledge of the Sirius star system reportedly far predated modern Western science (Kamalu 2018) and suffused the nonlinear notions of time and space that configured Afro-futurism from Sun Ra to Parliament, George Clinton's intergalactic ensemble through to the Detroit techno-pioneers. Mbembe draws attention to "a different kind of geometry out of which concepts of borders, powers, relations and separation derive" (4). This is a proposition orientated toward a Pan-Africanism built out of an alternative, non-Western set of coordinates, starting points, circulations, and interactions from elsewhere, movements of people, things, and ideas beyond the colonial and capitalist ordering of the world.

Both de Medeiros and Mbembe in their own profound ways set out a mode of thinking about living and being in the world that is Afro-centric in its conception, defined by a cosmopolitan disposition and characterized most notably through its nonlinearity. In returning to the purpose of this book— to think anew about the role of infrastructure in Africa's urbanization— the works of de Medeiros and Mbembe help make clear the need to shift beyond a singular notion of urban modernity. This book has attempted one such telling that has built out its understanding of this imperative to move beyond the linear and Eurocentric. It has done so through the proposition of the Infrastructural South. A way to move beyond dominant traditions of urban theory that produce problematic knowledge and representations of African cities.

WHAT IS THE INFRASTRUCTURAL SOUTH?

In chapters 3–8, I set out a series of techno-environments that establish the Infrastructural South as a material geography, that is, "dense networks of interwoven socio-spatial processes that are simultaneously local and global, human and physical, cultural and organic" (Heynen, Kaika, and Swyngedouw 2006, 2). I used analysis of techno-environments rather than infrastructure itself to denote the broader urban worlds that shape and are shaped by these systems beyond the purely technical. I developed a theoretical foundation in the opening two chapters by setting up the

10.1 Vodunaut by Nolwennlaureg is licensed under CC BY-SA 4.0.
https://creativecommons.org/licenses/by-sa/4.0/
https://commons.wikimedia.org/wiki/File:VODUNAUT,_Emo_de_Medeiros,_2021.jpg

Infrastructural South not just as a condition or geography but also as an epistemological position. If, as I argued, we require new critical ways to think about, research, and write about urban networks through a postcolonial and UPE perspective, this book has set out an approach that has a different starting point from the dominant traditions of social science research on/in/of Africa. The Infrastructural South is a basis through which to engage with these techno-environments and to help us think about a new set of infrastructural coordinates into the politicized presents and futures of urbanization. Chapter 9 demonstrated that the task of thinking anew generated by the idea of the Infrastructural South is not confined to those cities that through the infrastructures in operation have been ascribed as unmodern; it includes some urban spaces of the North where assumptions about a particular kind of networked or logistical city no longer hold.

Analysis focused on the operations, geographies, politics, and imperatives that make up a techno-environment. The assembled techno-environments are not a definitive account or meant as a full stop in advancing knowledge concerned with the role of infrastructure in urbanization, but rather they help to establish one possible way forward in thinking and working through the myriad of configurations and constellations at work. The need to bring into view a whole series of other techno-environments not covered in this book is of course pressing and constant, requiring multiple perspectives, voices, and contributions.[2]

I showed how the Infrastructural South that emerged from this investigation is more contradictory and complex than we would have expected if reading urban Africa through the abstracted, universalized model of urban modernity. The book's findings reject the representation of these conditions as merely *un*modern and playing infrastructural catch-up to the urban worlds of the North. In what follows, I reflect on how these techno-environments prompt us to revisit knowledge and assumptions. The aim is to establish a loose glossary through which to speak back to debates in urban theory. In doing so, I set out the knowledge and urban politics that come into view and those that might be necessary to open up new spaces of possibility amid intensifying urbanization.

OPERATIONS AND GEOGRAPHIES

The operations and geographies of the techno-environments producing the Infrastructural South are multiple, overlapping, and, in many regards, paradoxical. They include the wave of new cities and urban extensions being envisaged and constructed with increasing intensity, supported by surging finance and city models primarily from the East. The making of these enclaves as archipelagos of infrastructural security suggests African cities, or at least parts of them, as dynamic if segregated sites of new sustainable technologies, experiments, urban forms, and materialities, alongside growing wealth among certain sectors of the population. I examined the massive surge of finance into the speculative infrastructure corridors being proposed and constructed across the region. In doing so, I showed how these infrastructurally mediated plans and flows of finance involve the extensive, speculative restructuring of urban and regional space. They

Table 10.1 A glossary of the techno-environments

Techno-environment	Infrastructural operations	Infrastructural Geographies	Uneven geographies	Imperatives + openings
Enclave	High tech, infrasecurity, premium access	Securitized archipelagos	Eco-segregation	Democratizing experiments with new technologies, materials
Imposition	Breakdown of basic services, lack of repair + maintenance	Extended forms of urbanization	Uneven development + ongoing inequality	Decolonizing the material geographies of the city
Incremental	Self-built, peopled, fragile + ever shifting	Urbanization from "below" + beyond formal/informal	Survival driven, struggles for social reproduction	Building on prefigurative politics, heterogenous systems + popular economies
Corridor	Interurban connections + flows, speculative	Extensive restructuring of urban + regional space	Displacement, extractive, failure	Shifting away from extractive + neoliberal systems to social
Secondary	Series of ad hoc/uneven investments + substandard services	Infrastructural "catch-up" + urbanization "off the map"	Reliance on external actors + opportunities, loss of urban autonomy	Building platforms of experimentation + innovation, municipal control
Digital	Mobile tech, apps + devices, city as code	New digital spaces, layering of existing systems	New forms of extraction, control + surveillance	Building on efficiencies + popular economies
Predicament	Disinvestment, racialized austerity and breakdown	Decaying networks, peopled infrastructure, new spaces of extraction	Inequalities, struggles for social reproduction	Shifting away from networked city model as guiding logic/the North as center of global infrastructure power

do so through standardized, high-tech infrastructural configurations that attempt to (re)integrate Africa into global logistical networks and generate new interurban constellations as these materialities; expanded ports, new railway lines and roads, export and manufacturing zones, and associated technologies shift economic relations away from historic connections with the North. I considered the onset of a digital phase of infrastructure operations being layered across African cities and how new actors enmesh these technologies across existing operations. These interventions produce new geographies of urban space, defined, shaped, and governed through mobile technologies and accompanying code, algorithm, and app. This is proceeding in urban contexts that create a unique techno-environment of rapid experimentation and deployment. They bypass historic technologies (e.g., the landline) and are involved and integrated in the urbanization process from a much earlier stage than previous waves.

Taken together, these enclaves of sustainable, experimental technologies, the changing global economic landscape, new massive infrastructure corridors, and the advent of digital operations confront stubbornly persistent tropes of a technologically backward region. We might even argue that amid these reverberations, we find parts of urban Africa infrastructurally exceeding the disintegrating networks of postindustrial cities in the North such as Camden, New Jersey. What comes into view from these geographies and operations is a unique and diverse experience of infrastructural modernity that situates urban Africa at the forefront of the making of techno-environmental futures—a critical juncture to think about the role of infrastructure in the third wave of urbanization from a different vantage point to many of the popular representations associated with it.

Alongside these operations and geographies that challenge the idea of an infrastructurally challenged Africa, I outlined techno-environments that predicate ongoing struggles for survival and everyday social reproduction for vast numbers of people. Impositions of overlapping governance regimes produce an extended time/space of infrastructure that reinforces the racialized and class-based divisions and inscriptions of colonial and postcolonial eras in new ways. These imposition techno-environments materialize as breakdown of basic services and associated lack of repair and maintenance regimes. The inequalities of the urbanization process echo across multiple times and spaces, including as I demonstrated the unequal geographies of sanitation in Cape Town and disrupted electricity flows across Accra.

The response by urban dwellers to these impositions led me to explore the incremental techno-environments, which I characterized as self-built, peopled, and ever-shifting formal/informal operations and geographies. I examined the everyday struggles to connect to and sustain various circulations of energy in Ga Mashie and Namuwongo across hybrid, heterogeneous, and multiple systems of provision. These more than material arrangements were overlapping, fragile, and open to various types of pressure, whether economic, governmental, technical, or otherwise. The techno-environments that allow social reproduction under conditions of poverty and marginalization are also vulnerable to the effects of displacement, particularly through the impacts of state-capitalist projects promising better futures. New corridor-led urban and regional restructuring sometimes shatters the intricate and delicate social infrastructures of everyday life in ways that make the living of precarious situations even more challenging. And finally, I considered how these uneven geographies incorporate those secondary cities and towns, navigating uneven investment flows and substandard services that establish an infrastructural catch-up. This is a catch-up of both the urban service provisions of the abstract networked city model and the large, better fiscally supported capital and mega cities. These secondary techno-environments are often playing out "off the map" of policy makers and research, while reliance on external actors for infrastructure investment and opportunities limits urban autonomy and local capacity, fueling popular anger and uncertain urban futures.

In urban spaces characterized by these uneven operations and geographies, existing literature has often (mis)understood "modern infrastructure" as a particular set of technologies, imaginaries, operations, and geographies that are partial or absent. This should not be the case if we think more broadly. Rather, the notion of "modern infrastructure" can also be a container for this myriad of conditions, relations, processes, technologies, materialities, flows, and circulations that I bring together through the assembled techno-environments the book has set out. To be the "modern urban citizen" (Kooy and Bakker 2008, 376) in these contexts is not simply about who has access to networked services. For many, it is to navigate the irregular circulations of resource flows, the sometimes on, sometimes off electricity or water network, the knowledge of whom to get in touch with to fix, repair, or connect, and the collaborations and support networks that make life viable on and off various grids.

Modern infrastructure is that which is made and remade, broken and rebuilt, put together and taken apart by those for whom the promises of safe, reliable, and affordable urban service provisions and associated resource flows have never been fully extended by the state or market. What I wish to propose here is that whether the informal settlement or the high-tech enclave, we can find the techno-environments that compose urban modernity. If these networks are not reflective of the modern infrastructural ideal (Graham and Marvin 2001) and its modernist imaginaries (Jaglin 2015; Monstadt and Schramm 2017), it would be wrong to ascribe them the category of *un*modern. Rather, what I propose is to conceive of them as an extended or mutating form of modern configuration that operates outside, beyond, against, and sometimes even through the networked city model.

We should not read the techno-environments of urban modernity through the narrow meanings of modern, networked infrastructure, that is, a centralized sewerage network connected to flushing toilets or the always-on electricity networks with universal distribution. Rather, the urban modernity experienced in the Infrastructural South is one that is suffused by a distinct series of techno-environments—the extended time/space of impositions that reinforce historic inequalities, the emergent high-tech enclaves, the incremental, irregular systems of the popular neighborhood, the emerging digital layers of urban space, new corridors deployments, the infrastructural catch-ups of secondary cities and towns. We can be certain that there is no linearity in terms of modernity's urban histories, presents, and futures. Different spaces and places are technologically and environmentally distinct.

There is evidently no universal blueprint of what an infrastructural modernity is, even if we can be clearer about its critical role in understandings of modernization as denoting "specific processes such as the development of technological networks" (Gandy 2014, 4). I am suggesting, based on the explorations outlined in this book, that urban modernity cannot be teleological or universalized, or have an infrastructural end point as a particular centralized network. Rather, it is contextual, geographically differentiated, and sustained by a myriad of more than technical operations and geographies. These conditions force us to rethink our assumptions about the infrastructural dimensions of modernity and how we might navigate beyond them.

What became clear, given such findings, was that the historical techno-environments of modernity in cities such as London or Paris could not provide a blueprint for thinking through the role of infrastructure in global urbanization. A developmentalist, teleological way of thinking about infrastructure networks across cities is limiting. As I showed in chapter 2, Eurocentrism has built such a perspective in which the Western experience of technologies taming natures across urban space through particular infrastructural configurations became abstracted and universalized as a marker of progress. Therefore, existing frameworks of analysis could not account for the modern infrastructural geographies of Accra, Cape Town, Kampala, Mbale, Polokwane, or indeed other African towns and cities without ascribing these spaces as *un*modern, reproducing a problematic and wider history of representation. My intention was therefore to shift beyond the false binary of the modern metropole and the unmodern periphery predicated on this Eurocentrism and critical to the production of a hierarchy between global cities (J Robinson 2006) into the variegated but relational geographies of urban modernity and its multiple, heterogenous infrastructure. Given the need to reject a Western-centered set of narratives about the techno-environments that produce the urban condition, I proposed a revised approach to infrastructures and urbanization.

MUTATING MODERNITY

A key proposition that I think is required, given the operations and geographies of the Infrastructural South, is to reorient the basis of our understanding of urban modernity and specifically its techno-environments through the idea of a mutating modernity. "Mutating," according to the Oxford English Dictionary, implies a "change in form or nature" and helps us to establish better the basis for bringing into view the ever-shifting techno-environments of modernity. The term destabilizes binaries, assumptions, and ways of doing research and undertaking analysis. It draws attention to changes to the "normal" and "abnormal" process of reproduction. I do not mean it to highlight that we differentiate only some places and spaces from an abstracted experience, or that this is a world with a multiplicity of modernities (Eisenstadt 2000). Mutating modernity is a way of thinking about the UPE of infrastructural urbanization in which all techno-environments are always contingent, differentiated, and relational across the time/space of urban worlds. This is a mode of thinking about the technological and environmental dimensions of urban space as mutating from an abstracted form of urban modernity, and a particular experience of urbanization, highly connected to certain configrations of infrastructure and the networked city model. It draws attention to thinking about the making of modern, urban Africa in new politicized ways.

Mutating modernity can help to destabilize assumptions and Western knowledge traditions in ways that open up new analytical potential and pathways in thinking through knowledge concerned with the global geographies of infrastructure and which offer the basis upon which I have elaborated the Infrastructural South in the book. This is a foundation through which to think anew about the techno-environments of the modern world, about a mutating modernity in the rapidly growing towns and cities of Africa and many elsewheres. This is a counter to the logic of linearity and singular history as told from and radiating out of the West that rejects a "global north to be the unique legislator of a planetary modernity" (Chambers 2017, 20) and an opening into the Infrastructural South. And mutating modernity is as relevant in thinking about the techno-environments of Europe and America as Africa because the abstracted model, which has flattened multiple, heterogenous urban histories is certainly not reflective of the contemporary urban political ecologies at work in cities such as Camden, New Jersey, lead-poisoned Flint (Pulido 2016; Ranganathan 2016), or Detroit (Gaber et al. 2021). It prompts a focus on the relational ways the metropole was made modern through the underdevelopment, exploitation, and extractions visited on the South, but also the predicaments now facing the metropole.

UNEVEN GEOGRAPHIES OF THE INFRASTRUCTURAL SOUTH

UPE has long drawn attention to the uneven geographies and inequalities that shape the urbanization process (Swyngedouw and Heynen 2003;

10.2 We want water graffiti in Kibera, Nairobi.

Tzaninis et al. 2021). It has done so in ways that span from the geological (Dawson 2021) to the embodied (Doshi 2017; Truelove 2019), demonstrating how multiple sites, spaces, materialities, and processes are entwined with the technological and environmental. Throughout this book, I have loosely drawn on these critical traditions of UPE to set out the production of uneven geographies that are so integral to the Infrastructural South. I set out to consider who shapes these and how and, concurrently, the ways urban dwellers might experience these techno-environments unequally. And out of these uneven geographies, an urban politics of the Infrastructural South comes into view.

In thinking through the enclaves of Accra, I argued that the emergence of these archipelagos as new cities or urban extensions was predicating a new form of spatial division that I defined as eco-segregation. In such contexts, the elite can increasingly pay for an experience of infrastructural security with always-on services and increasingly "sustainable" technologies, materialities, and ways of living. Those without the finance to claim a part of this pixilated modernity are left to navigate the underinvested infrastructure networks of the wider city. With climate change and other environmental crises intensifying, these enclaves will mark a clear, ugly division in urban society—a demarcation that echoes the segregated techno-environments of the colonial era in new uneven and unequal ways.

Focus on imposition techno-environments made clear how extended urbanization through which infrastructure is (re)produced is critical in the making of uneven geographies. I set out how inscriptions of historical eras of governance inside and outside the time and space of the city configure the Infrastructural South. I explored the colonial and neocolonial logics that shape(d) broader urban and regional geographies of contemporary experiences of electricity in Accra. These impositions become central to explanations of the production of network disruption and the social costs of *dumsor* for those in the city without the resources to afford backup power systems. In Cape Town, I showed how the inscriptions of (settler) colonial and apartheid logics continue to shape uneven development through the racialized operation and experience of sanitation in Black urban spaces such as PJS Section. These are forms of governing that, despite being cast aside decades ago, remain a powerful structuring force in and on the apartheid infrastructure of the city.

The chapter on incremental techno-environments focused on and outlined some of the multiple struggles relating to access of safe, affordable, and reliable flows of electricity in the city. As others in disciplines such as urban geography and anthropology have explained (Baptista 2015; Castán Broto 2019; Guma 2019), there are a myriad of ways in which people have to interact with energy networks to power everyday life, given the less than universal networked geographies across African towns and cities. These practices, collaborations, strategies, and tactics are both reflective of, and further reinforce, uneven geographies of distribution, access, and usage. A material and political engagement with infrastructure draws attention to how state and capital can deem urban populations as surplus and their infrastructural needs ignored. I traced these intersections through techno-environmental struggles of social reproduction and the securing of resource circulations and flows across the city. These modes of infrastructural survival are a central feature of urbanization in African towns and cities (De Boeck 2015). They point to the ways in which intersections of people and infrastructure "rework embodied labor, differentiated citizenship, and socio-ecological relations" (Ramakrishnan, O'Reilly, and Budds 2021, 669). The resulting uneven techno-environments are often more expensive that the urban services delivered by the state and can be dangerous for those who are forced to become engineers to create new configurations. They may be unsafe in other ways such as contaminated water, electrification, or risk of disease transmission, and they can leave residents exposed to unequal power relations with informal providers (Boakye-Ansah, Schwartz, and Zwarteveen 2019; Swyngedouw 2004). And the likelihood of breakdown through malfunction or disrepair across these techno-environments remains high, as do the risks of this urbanization from below being ruled or classified as illegal and destroyed by the state or utility companies. Clearly, incremental techno-environments are the most visceral and visible infrastructural form of inequality.

The new era of proliferating infrastructure corridors promises economic growth and interurban and national connections to speed up various types of circulation, from finance to natural resources, people to ICT. However, as I highlighted in Namuwongo, establishing the urban spaces for these massive infrastructures can cause displacement for resource-poor communities living precariously on contested lands claimed by the state. With the Central Corridor, I showed how the fragile infrastructures of social reproduction in the neighborhood risked being fractured as the Uganda Railways Corporation imposed its claims of these spaces. And the vast investments required to make these geographically extended systems are always full of risk. Policy makers and politicians face prospects of failure through various means: debt defaulting, underperformance, breakdown of diplomatic relations, and so forth. As I noted in Port Bell, with the arrival of the timber from eastern DRC, infrastructure corridors are also being deployed in the making of new "frontiers of extractive capital accumulation" (Lesutis 2020, 600) across hinterlands. Extractive geographies have historically relied upon urban situated infrastructure as nodes within broader flows and configurations of global capitalism in ways that have and likely will continue to underdevelop Africa (Rodney 1972) while exposing populations to harm, insecurity, and uncertainty.

I made clear the uneven geographies between the large metropolitan regions and small cities/towns in Africa across the secondary techno-environments of Polokwane and Mbale. I showed the infrastructural catch-up facing these municipalities as they struggled to secure necessary financing and investment. The lack of fiscal capacity has generated an openness to accept opportunities for new flows of finance, whatever the costs. This has led to reliance on external actors for grant funding and a subsequent lack of autonomy to make local decisions about infrastructural futures. I explored the deployment of PPMs for electricity and responses to water-scarcity in Polokwane and a waste project financed and designed by the World Bank in Mbale. In both cases, contestation and resistance from local communities, workers, politicians, and political parties are intimately tied to the infrastructural catch-up. Uncertainty about how and when the catch-up will be achieved and through what means pervades these secondary techno-environments.

Despite the multiple promises that follow the deployment of digital technologies, these relatively new attempts at layering of urban space through code, algorithm, and app are reinforcing both existing and new uneven geographies. Novel forms of (digitally enabled) extraction were evident through the global expansion of Uber into African cities such as Kampala, reflective of longer histories on the continent and of how the digital is producing and new accumulation regimes. Meanwhile, the capacity of the state to exert digital control and surveillance across cities during crisis moments such as the Ugandan election points to how democratic space continues to be closed down in new ways. These new digital layers provide authoritarian governments with new tools to close down and repress the potential for urban populations to resist and contest their lived conditions.

Given these multiple, connected uneven geographies of the Infrastructural South and the injustices and inequalities being produced, it is clear that some form of rupture from what has gone before is required in the future. This must be a decisive shift away from the contemporary ordering of techno-environments through which the Infrastructural South has been produced. Begüm Adalet (2022) wrote about "infrastructures of decolonization" as a way to rebuild urban worlds. He argued that the ideas of Fanon are critical to this task because they are "attentive to the ways in which colonialism operates by reproducing certain forms of spatial organization across multiple scales. He proposes a type of anticolonial worldmaking" (23). In effect, this world making means moving away from how infrastructure has been planned and operated. Doing so may help to remake the geography of the Infrastructural South. This is what Shilliam (2016, 425), again drawing on Fanon, argues as the need to "dismantle the architecture of the master and redeem a space for living other-wise."

FROM CRITIQUE TO PROPOSITION

To think of a popular infrastructure as a possible future, trajectory, or journey from/with/for the Infrastructural South is to embrace a "propositionality to advance southern urban infrastructure debates towards more anticipatory forms of scholarship" (Baptista and Cirolia 2022, 1), which "not only critiques . . . but also experiments, imagines, and inspires." Doing so directs focus on the critical role of techno-environments in the Fanonian "world making" that Adalet (2022) points toward. The idea of popular infrastructure shifts the Infrastructural South from an epistemology focused solely on problematization and critique toward a basis for thinking about world making through infrastructure. Foremost, I contend we must root developing notions of a popular infrastructure in commitments to democratize and socialize power and ownership over urban networks, its primary task being to address techno-environmental injustices. Popular infrastructure opens up a space to join the conversation about how towns and cities navigate the uncertain futures of the third wave of urbanization. It is a commitment to finding, deliberating, and reworking forward-looking ideas that begin from a critical knowledge of the techno-environments that produce the Infrastructural South.

Debates and ideas concerned with popular urbanization have a long history, especially across Latin America, with writers such as Emilio Duhau (1992, 48) thinking about how the city is built from the ground up, in which "a very large number of families solve their housing problem by acquiring land under irregular conditions and self-producing their habitat. This is what we call *urbanización popular*." The term has then been used more recently "to describe a specific urbanization process based on collective initiatives, self-organization and the activities of inhabitants" (Streule et al. 2020, 652). And AbdouMaliq Simone (2022, n.d.) has written about how popular economies operating across urban worlds point to ways for "how people can be brought together more effectively and judiciously to cooperate with each other, share their ideas and experience, and view themselves as operating in the city together."

I propose to extend this idea of the popular both into the infrastructural realm specifically and to incorporate a broader range of actors into the process, including the state itself. This is an invitation to place infrastructure in the service of attempts to weave "together whatever is at hand but collaborate with each other in the pooling of resources and opportunities" (Simone 2020, 612). Such collaborations would exceed popular if poor neighborhoods and span across the city and beyond. It is a disposition that can bind together municipal bureacrats with social movements, utility companies with young entrepreneurs in new forms of association unleashing collective action that can propose and build an alternative Infrastructural South.

A research agenda for this popular infrastructure is shaped by a political orientation toward the remaking (or world making) techno-environments. It embraces a mutating modernity, breaking away from what has gone before as a rigid model to be followed. In what follows, I set out an initial six key characteristics of this popular infrastructure that articulate possibilities and potential, pointing toward how scholars may engage broader efforts, processes of knowledge production, and struggles for a better world. These are: massive public works schemes, embracing heterogeneity and hybridity, new common forms of ownership, platforms of experimentation, equitable financing arrangements, and a reinvigorated Pan-African political imaginary.

MASSIVE PUBLIC WORKS SCHEMES

Current financial, technical, and governance arrangements bring together "interconnected economic, political, social and ecological processes that together form highly uneven urban socio-physical landscapes" (Heynen, Kaika, and Swyngedouw 2006, 16). These conditions cannot rapidly and at scale deliver the infrastructure required across urban Africa over the next few decades. A decisive shift to a popular infrastructure then requires a massive, transformational program of public works. I emphasize the public in this program for two reasons. The first reason is to signify a definitive break from the neoliberal governance that has accompanied the governing of infrastructure in recent decades. Whether full-on privatization, concessions, or PPPs, the World Bank and the IMF have pushed through market-based logics of infrastructure that cannot be the means for delivering and operating fair, safe, and universally accessible infrastructure for social reproduction. The critique of this World Bank and IMF ideology is widespread, finding expression in popular music such as Seun Kuti's 2014 putdown, "IMF"—the time to move on is now.

Second, the focus on the public acknowledges that without the direct involvement of the state, whether national governments, local municipalities, or state-run utilities, the remaking of existing (and unequal) techno-environments cannot be achieved. This is an acknowledgment of the experience, skills, expertise, and capacity contained within these entities. However, as I will set out below, the existing hybridity and heterogeneity of the Infrastructural South opens up the possibilities for a different role for the state as enabler or facilitator.

This would open up opportunities for multiple forms of collaboration and collective management in the delivery of this massive works program, from municipalities and cooperatives to community groups and other types of social organization. And the labor requirements of such a rapid implementation of public works would help to address the persistent and intensifying crisis of underemployment of youth in African countries such as Ghana (more than 50 percent) and Uganda (more than 60 percent). Indeed, Africa's working-age population is expected to grow by nearly half a billion by 2035, but with only 100 million estimated jobs. For these new workers, publicly controlled regimes of maintenance and repair that build out from and sustain into the future the massive public works schemes, properly financed,[3] would provide ongoing forms of secure, decent, and safe employment.

EMBRACING HETEROGENEITY AND HYBRIDITY

A popular infrastructure has to embrace the hybridity and heterogeneity that is already in operation in the Infrastructural South. This has to be a definitive shift by planners, policy makers, and others beyond a focus on centralized systems and a particular modern imaginary that this book

10.3 Laborers taking a rest in Mbale.

has set out to critique. Under these logics and assumptions, authorities cannot deliver a massive works program aiming to provide safe, universal, and affordable infrastructure services at a rapid pace. The networked city model is limited to thinking about Western cities of the nineteenth and early twentieth centuries, but certainly not the twenty-first century. As I have argued this model is not the only infrastructural modernity at play in the city. Embracing the multiple models that come into view through a focus on mutating modernity opens up new possibilities for delivering vital infrastructure. It provides a basis on which to shift decisively beyond the networked city model as the guiding planning logic or to view "modern" infrastructure services as only one particular configuration, imaginary, and set of operations. Rather, this is a commitment to experimenting, deploying, and upscaling multiple and overlapping configurations that can deliver at pace the basic urban service provisions required.

This is a dispensation that in part amplifies, standardizes, regularizes, and legalizes the hybrid and heterogenous infrastructure already in operation in many spaces because "urban politics basically has to find ways

10.4 A National Slum Dweller Federation–run toilet in Mbale.

of working with what exists" (Simone and Pieterse 2018, 89), whether
a social enterprise implementing new hand pump "gulper" technologies to empty toilets beyond the sewerage network (Nakyagaba et al.
2021), a community-run energy system providing off-grid solar generation (Mugisha et al. 2021), households producing their own energy bri-
quettes for cooking or Slum Dweller Federation–run toilet blocks (Burra,
Patel, and Kerr 2003). The potentials are almost endless and, in many
cases, already being actualized. This heterogeneity renders an alternative infrastructural ordering of the third wave of urbanization. As I have
argued with colleagues thinking through these alternative infrastructural operations (Lawhon et al. 2018, 730), "At an even wider scale, these
HICs could nurture a wider urban economy, providing both the context
for localised designs to emerge, and to be translated, or exported to be
used elsewhere, inspiring potentially transformative and more sustainable regional urban futures." The plethora of non-networked infrastructure
in operation as new forms of decentralized, autonomous, and localized
systems are scalable in ways that move decisively beyond the logics of
centrally planned large-scale technical networks but deliver the same
standards of urban services.

State-run utilities cannot monopolize repair and maintenance regimes under such conditions. Instead, an array of actors, already deeply involved in the hybrid and heterogenous systems being operated, should complement the role of the state and utilities. Think of the self-employed phone fixers in Kampala, or how the SJC/Ndifuna helped to establish a digital intermediation to enhance the maintenance of toilets in Khayelitsha. In Maputo, for electricians working across the formal and informal network, it requires "embodied assessment of what works and does not work in the re-assemblage of material objects, technological devices, and the complex socio-economic conditions of the urban environments they are a part of" (Baptista 2019, 516). Much like those working to maintain and repair the energy systems in Namuwongo and Ga Mashie, these skills, knowledge, and adaptive capacities are part of the operations of a popular infrastructure already. What is required is recognition and legitimation of such peopled infrastructures and the safety standards and financial rewards that are all too often missing from such roles.

COMMON FORMS OF OWNERSHIP

Third, a popular infrastructure involves common forms of ownership within and beyond the state. Primarily, such infrastructure commoning can democratize and incorporate the various popular economies at work in the heterogeneous and hybrid operations of infrastructure across urban Africa. Silvia Federici (2019, 719) posed the question, "What does it mean to speak of the commons as a mode of production and a principle of social organization in the present phase of capitalist development?" Thinking this through infrastructurally is vital and involves how these systems "could be more democratically, according to principles of collective control, participative decision-making, and a fair distribution of benefits" (Becker and Naumann 2017, 4). Primarily, I would contend, the answers revolve around strategies to hold together both the massive works programs delivered by the state and the making and operation of infrastructure from below in "a radical new conception of public ownership" (Cumbers 2012, 256). This is a commitment to build upon the ingenuity and knowledge of those people and communities already operating these urban networks to democratize, decentralize, and make popular the economies of infrastructure. This economic repurposing toward collective good and public value over private profit can best be delivered by infrastructural actors through establishing ways to open up ownership, control, oversight, and management of various infrastructure across different scales of operation in and beyond the city.

It means to recognize how NGOs, congregations, cooperatives, schools, and social movements are already or could be integrated into the various types of infrastructure operation, whether in production/generation, distribution, consumption, or management. There has been little direct scholarship on thinking through infrastructure as commons in urban Africa, despite much work on various anti-privatization campaigns and struggles and calls for decommodifying various resources flows such as water (Bond 2013; Narsiah 2011; Robins 2019). However, there are substantial debates and ideas that run across urban studies (Gidwani and Baviskar 2011) including work on claims for urban space by squatter movements in South Africa (Pithouse 2014) and architecture and planning practice in Ethiopian cities (Charitonidou 2021). These stories, inspirations, and struggles from urban Africa can help open up new ideas about how to think about and initiate common forms of ownership.

The type of common ownership will vary. Some infrastructures such as energy generation or infrastructure corridors might require ownership focused at the national or even international scale. This is in order to undertake strategic, long-term planning, particularly regarding rising energy demands and the growing specter of climate change, alongside often significant financial commitments. State ownership requires reversing the privatizations that have emerged in recent decades, or halting privatization processes that have not been effective (e.g., Appiah-Kubi 2001 on Ghana's experience). And for this (re)nationalization to be a type of genuine infrastructural commoning, it must embed democratic accountability. This could include boards made up of civil society groups, trade unions, and NGOs providing scrutiny over operations and investments and how these infrastructural entities follow strict social and environmental criteria. Large, democratic, state-owned enterprises offer the scaffold for more diverse forms of collective ownership at a series of different scales.

New and varied forms of municipal, community, and cooperative ownership across towns and cities will govern and control infrastructure. Municipalization and further devolution of ownership would include similar models of accountability and scrutiny, mirroring the state-owned enterprises detailed above, and would offer opportunities to deepen and broaden accountability and the sharing of benefits. The potential that emerges from this municipal ownership includes opportunities to develop local procurement practices that resist extractive practices through stimulating urban economies, allowing money to circulate among users and communities. Municipal ownership can open up new forms of experimentation and innovation that build heterogenous systems suited to particular topographies, conditions, economic circumstances, and people's demands.

Democratic member control, equality, and solidarity characterize cooperatives. They have a long history in many African countries, particularly during the modernization era of the postindependence years (Hamer 1981). They range from those owned by members or broader communities through to worker-owned forms, and they offer interesting ways to think about how common forms of ownership could spill out across the operation of popular infrastructure. Massive expansion of cooperative forms of ownership across the infrastructure realm would require various types of support to initiate finance for workers' buyouts, comprehensive expertise for cooperative development, shifting legal provisions, procurement contracting to prioritize cooperative enterprises, and so forth.

PLATFORMS OF EXPERIMENTATION

The making of a popular infrastructure requires establishing platforms of experimentation that can push forward just and sustainable technical, social, financial, material, and political ideas into new operations, forms, and features. Experiments proceed through pilot projects, innovations in collaborative and social organization, municipal and community-led initiatives, and novel technologies (Bulkeley and Castán Broto 2013). This is a prefigurative political disposition that requires "a commitment to experimentation, unlearning, shedding habits of thought and always figuring another angle despite the odds" (Simone and Pieterse 2018, 153) and which is represented so well by the artworks of Emo de Medeiros with which I began this chapter.

In Cape Town, I worked with the municipality to evaluate the results of a retrofit experiment with government-built RDP houses that had been badly (and cheaply) built, leaving households with various problems with heat, energy, and health. The pilot, funded by an international donor, took the form of retrofitting insulated ceilings into 250 houses in Mamre, north of the city. The aims were to improve thermal efficiency so that households would be warmer, spending less money on heating and having improved health outcomes, and to use local labor to undertake the program. The results that we presented in a report (Phillips, Silver, and Rowswell 2011) were encouraging, showing how this small intervention had a dramatic impact on households across a range of metrics. The City of Cape Town used these findings to make the case for further investment that was secured from the South African government's Green Fund, managed by the Development Bank of Southern Africa. What started as a small-scale experiment by a bunch of inspiring officials willing to try something new had snowballed. It led to the municipality being able to secure funds to invest ZAR 133 million (USD 8.5 million) to retrofit another 8,000 RDP houses and create 800 'green' jobs in the process.

In Ghana, I spoke to Renne C. Neblett at the Kokrobitey Institute outside of Accra, whose experimentation around Afro-centered sustainable technologies and infrastructures, built form, and architecture offered a very different vision of the future from nearby concrete enclaves. I asked Renne about the suburban style housing that characterized much of Accra and often came accompanied by air-conditioning systems. What she said has a much wider resonance to think about popular infrastructure in African cities. Renne argued:

There is no reason to copy Western building techniques except to show prosperity, no reason to build in concrete. It's the antithesis of what should be getting built. I want to be clear about traditional building and traditional building techniques that can be appropriated for modern design building. I think it's an important distinction. No one should live in the past. I think what we grapple with is what was progressive in the past and what should be left behind and what we try to take forward is traditional technology and intelligence about the climate and turn it into something modern and relevant.

This experimentation at Kokrobitey brought together centuries of African knowledge, expertise, and traditions that have produced climate-resilient building techniques and materials into new architectural and infrastructural expressions for the contemporary city. This was an embracing of both

10.5 The Kokrobitey Institute.

past and future to assemble new hybrid directions, pathways, and journeys that might offer a template to shift beyond Western-centric ways of doing things.

REPARATIONS AS EQUITABLE FINANCING ARRANGEMENTS

What financial architectures might support design and deployment, operation and maintenance of a popular infrastructure? There is a vast web of financing and financial actors entangled across the infrastructure of urban Africa. The first of these is through growing fiscal capacity within Africa itself. National governments are increasingly able to access support from region-wide financial organizations such as the African Development Bank Group and those operating in particular parts of the continent such as the Development Bank of Southern Africa. And institutions such as the African Union and Economic Commission for Africa have accelerated new financial mechanisms as they acknowledge the need to build sustainable infrastructure rapidly (Hyman and Pieterse 2017). Second, there are high-profile, new investment flows emanating from China and from banks such as Exim (Adunbi and Butt 2019) through the large-scale, loan-based infrastructure finance connected to the BRI (ZiroMwatela and Changfeng 2016). Cumulative loans since 2000 now make up more than USD 140 billion as China becomes the largest bilateral lender in the region (Lokanathan 2020). Third, new global circulations of speculative investment through what has been termed the "infrastructure–finance–real estate nexus" (Goodfellow 2020) have enabled the building of private developments such as those we encountered in chapter 3. This financing has become a mechanism for new private forms of infrastructural solution.

Collectively, these financial geographies have dramatically and rapidly reshaped how actors have funded urban-sited infrastructure in recent years, even if the final forms remain caught in problematic logics, anticipations, and assumptions. The issue remains not just whether these often financialized, market-driven mechanisms provide investment for public infrastructure but also whether these investments are enough to address urbanization imperatives, even under current logics and assumptions. Given the finances that would be involved, hundreds of billions of dollars per year, I would suggest not. Rather, a new set of subnational financial arrangements and a new global financial settlement must be fought for.

I would contend that the requirement to transform fiscal architectures to support a popular infrastructure requires a massive upheaval in finance that radically breaks from what has come before and is based on a reparative politics. This means focusing firmly on the global system of racial capitalism that led to the underdevelopment of Africa and which continues to the present day. As Moshood Kashimawo Olawale Abiola (1992, 910) questioned, "Who knows what path Africa's social development would have taken if our great centers of civilization had not been razed in search of human cargo?" Such huge global financial inequalities have been calculated. One study suggested that African countries "lose USD 203 billion through factors including tax avoidance, debt payments and resource extraction" and an "annual net financial deficit of over USD 40 billion" once loans, aid, and remittances are factored in (Curtis and Jones 2017). This financial imbalance is significant, and it helps to explain the lack of finance to support the infrastructures required for Africa's urbanization, but if anything, it is an underestimation, failing to account for a series of historical-contemporary factors that perpetuate the underdevelopment of Africa, none more so than slavery and climate change.

Calls for reparations for slavery and colonial extractions and the centuries-long impact on Africa have a long, complex history (Howard-Hassmann 2018). Although financial estimates vary, these could amount to USD 100 trillion (Osabu-Kle 2000), which compares to more than USD 700 billion of external debt currently held by Africa (World Bank 2022). The Group of Eminent Persons was set up in the 1990s by the Organization of African Unity. It notably drew on the underdevelopment thesis of Walter Rodney (1972). The Group pushed for a conception of reparations that would give recognition to the claim by Mazrui (1999) and others that Africa developed the West, which the opening chapters of this book considered through the specific focus on the relational, uneven making of urban modernity.

The last significant Pan-African discussions on reparations were held in Durban at the UN World Conference against Racial Discrimination, Xeno-phobia and Related Intolerance in 2001. The outcome disappointed many reparation campaigners after a diluted final statement. While there has been little recent political energy spent on reparative claims in institutions such as the African Union, the issue remains a key touchstone for demands for global justice. For instance, in 2013, the Caricom Reparations Commission, set up by heads of state in the Caribbean, aimed at "repairing the damage inflicted by slavery and racism" (Franklin 2013, 365). The type of repara-tions proposed included debt cancellation and interestingly, when thinking about the Infrastructural South, technology transfer (Bicknell-Hersco 2020), given colonies were "denied participation in Europe's industrialization process" and were "confined to the role of producer and exporter of raw materials" (Caricom, online).

Another claim for reparations for Africa's historic underdevelopment and recognition of the massive, interlinked crisis across cities has been ori-entated around climate change and acknowledgment of the climate debt by the North (Adow 2020; Bond 2010). The continent has contributed little in the way of historical, global carbon dioxide emissions. The case for repara-tions for this historical, colonial-capitalist pollution of the atmosphere is striking. Climate change is intersecting with the infrastructures of the third wave of urbanization in ways that demand rapid adaptation and mitigation across urban space (Parnell and Walawege 2011). The time for polluters to acknowledge these climate debts and for global institutions to provide the finance mechanisms for popular infrastructure to address the impending climate crisis is now and must be rapid and significant. This should not be in the form of more loans or debt or with any so-called conditionalities. As Ajl (2021, 11) argued in relation to the People's Agreement of Cochabamba, such decolonial praxis, built on acknowledgment of climate debts, might open up "planks of a Southern platform for ecological revolution."

A PAN-AFRICAN POLITICS OF POPULAR INFRASTRUCTURE

Developing new public works schemes, common forms of ownership, embracing heterogeneity, fighting for equitable financing arrangements, and establishing new platforms for experimentation are all vital in the making of popular infrastructure. These imperatives would also require a radical transformation of the political economies of urbanization and the global economic system. If this seems unlikely, given centuries of colonial-capitalist ordering, it is not impossible and can begin from African states and peoples finding inspiration from the forward-looking modernization era of Afro-socialism and reinventing it for the contemporary moment. This was a Pan-Africanism that combined the struggle for equality, a commitment to anti-imperialism, and shared economic power with traditional African-based ways of knowing the world, of being a community, and of caring for one another (Akyeampong 2018; Mohan 1966).

This politics built on traditions in which "it gave members of the society a secure and relatively adequate livelihood, and it gave them a full opportunity to share in the making of the conditions upon which their happiness depend" (Mboya 1963, 19). As Julius Nyerere (1964, 164) argued, "nobody starved, whether of food or human dignity because he lacked personal wealth; he could depend on the wealth possessed by the community of which he was a member. That was socialism. That is socialism." If the programs of Nkrumah, Nyerere, and others were predicated on modernization, what comes next will be fundamentally different, given the world we now live in and the techno-environments we inhabit.

Collectively, these routes, pathways, and journeys toward a popular infrastructure are a commitment to a decolonized city and urban (knowledge) politics required to address the challenges and imperatives of infrastructure in the making of the third wave of urbanization, from reparative claims to architectural experiments to new common forms of ownership, assembling traditions and futures that mutate into new openings. As Thomas Sankara (1988) argued, "We must dare to invent the future" and that such boldness "comes from nonconformity, the courage to turn your back on the old formulas, the courage to invent the future." The next decades will tell us whether such possibilities can be brought forth, whether urban Africa will find new ways to be governed and managed, lived and experienced in which, in the words of Nkrumah (1997), "We face neither East nor West; We face forward." Research on the geographies of infrastructure and the techno-environments that surround and suffuse these networks are an essential foundation in supporting these journeys. A new, inspiring generation of African scholars are leading the way and may in their future journeys also offer new ideas for the crumbling systems of many European and American cities. They will guide us all in thinking critically and propositionally about how we undertake our research and politics. The Infrastructural South I have set out in this book and the popular infrastructure of this chapter give, I hope, one partial route forward in the critical task of knowing and addressing the imperatives of these uncertain futures.

NOTES

1 INTRODUCTION: INTO THE INFRA-FUTURE

1. This was achieved in 1961 in Latin America and in 2020 in Asia, including China in 2011 and Southeast Asia in 2020.

2. Apart from Oceania, which was already majority urban population but has a very low total population compared to other world regions.

3. While the Reconstruction and Development Programme as the implementation of investment into housing and infrastructure finished in the 1990s, the generic countrywide model/typology of house generated during the process has come to dominate the government-financed house-building process.

2 URBAN MODERNITY AND THE AFRICAN CITY

1. Faust is the protagonist of a classic German legend. He is a scholar who is highly successful yet dissatisfied with his life, and his dissatisfaction leads him to make a pact with the Devil, exchanging his soul for unlimited knowledge and worldly pleasures.

3 INFRASTRUCTURAL SECURITY TO ECO-SEGREGATION

1. While other researchers have estimated a greater number of new cities (e.g., van Noorloos and Kloosterboer [2018] suggest up to seventy existing and planned African new cities), the collection of data focused only on projects either in construction or likely to proceed, and developments that were fully separated from the broader city, and were larger than a suburban, gated community or state-led housing such as in Luanda.

2. There is a recent history of development projects being given tax exemptions in Accra, such as the Pullman Accra Airport City Hotel, which was granted more than USD 20 million of exemptions in 2020 by President Akufo-Addo.

4 BETWEEN SURVIVAL AND THE PREFIGURATIVE

1. This relatively recent work on the incremental in postcolonial urban studies has drawn on a longer tradition of scholarly attention in development studies research across regions including Latin America, Southeast Asia, and sub-Saharan Africa. This broad and extensive literature has particularly focused on housing) that has helped to shape development discourse and practice, especially around notions of incremental upgrading of existing informal settlements and associated strategies of poverty reduction.

2. Powering Namuwongo is available to watch at https://vimeo.com/158609915.

5 THE EXTENDED TIME/SPACE OF INFRASTRUCTURE

1. The DA is the main opposition party in South Africa and is in control of the City of Cape Town and the Western Cape province. It was formed in 2000 with a history as the white parliamentary opposition to the National Party.

2. South Africa's Gini coefficient ranges from about 0.660 to 0.696. The Gini coefficient is the measure of income inequality, ranging from 0 to 1. 0 is a perfectly equal society, and a value of 1 represents a perfectly unequal society.

3. The Eastern Cape was purposefully underdeveloped by the apartheid regime, leaving a legacy of poor employment, economic infrastructure, and basic services that have forced many people in the post-apartheid era to leave their homes in search of opportunity.

6 PROMISES OF DEVELOPMENT, EXPERIENCES OF DISPLACEMENT

1. Other spokes of the Central Corridor will extend from Tanzania to Rwanda and Burundi.

8 DIGITAL DISRUPTIONS FROM "ABOVE" AND "BELOW"

1. The International Labor Organization reported that "the youth labor underutilization rate in Uganda was high at 67.9 per cent in 2015" (ILO 2015). With one of the youngest populations on the planet, with a median age of 15.8 years, which is only lower in Niger, the need to find livelihood opportunities remains one of the biggest challenges for Uganda's urban youth.

2. This included the house arrest of opposition leader Kizza Besigye copied into the 2021 election with Bobbi Wine, through to the prohibiting of public meetings by opposition activists in the run-up to the election, to the attack on and siege of the FDC offices, the circulation of people involved in politics and against the National Resistance Movement (NRM), known by many Ugandans as the "National Robbery Movement."

9   POSTCOLONIAL PRESENTS IN THE METROPOLE

1. Abu Dhabi United Group are involved in a real-estate venture with the local municipality, Manchester City Council, in a series of companies that together constitute the Manchester Life Partnership.

2. See also David Harvey (2007) on a longer history of neoliberalism, being pioneered in Chile.

3. Privatization took place from 1999 in a twenty-year, USD 215-million contract awarded to Bechtel Group/United Utilities, and a revised contract in 2016 to American Water.

10   TOWARD A POPULAR INFRASTRUCTURE

1. One of their legends is based around a visit thousands of years ago by alien species called Nommos from the Sirius star. The Dogon were told by the Nommos of a sister star, Sirius B, which was only discovered by scientists in 1970.

2. For example, the pathogenic techno-environments that shape circulations and experiences of long-standing diseases across African cities such as malaria or tuberculosis (alongside newer pandemics such as COVID-19 and Ebola) or the robotic techno-environments that are enabled through massive advances in computing powers in which new automated and robotic technologies usher in an age of drones and automated work.

3. In South Africa, the EPWP is one of government's key programs aimed at providing poverty and income relief for the unemployed and has often involved various forms of infrastructure repair and maintenance. However, the length of employment has been limited to twelve months, and remuneration is low at ZAR 12.75 (less than a dollar an hour). Despite these issues, some participants have gone on to full-time employment, with the City of Cape Town suggesting it has permanently employed 3,600 workers over the years through the scheme. The EPWP then shows both the potential of massive public works schemes and the necessity for them to be properly funded.

REFERENCES

Aalders, Johannes Theodor. 2021. "Building on the Ruins of Empire: The Uganda Railway and the LAPSSET Corridor in Kenya." *Third World Quarterly* 42 (5): 996–1013.

Abani, Chris. 2005. *Graceland*. New York: Farrar, Straus and Giroux.

Abbou, Tahar. 2016. "Mansa Musa's Journey to Mecca and Its Impact on Western Sudan." Accessed January 20, 2022. https://www.researchgate.net/publication/34343 1925_Mansa_Musa's_Journey_to_Mecca_and_Its_Impact_on_Western_Sudan.

Abiola, Moshood Kashimawo Olawale. 1992. "Why Reparations?" *West Africa* 1 (7): 910–911.

Abu-Lughod, Janet L. 1989. *Before European Hegemony: The World System AD 1250–1350*. Oxford: Oxford University Press.

Acey, Charisma. 2012. "Forbidden Waters: Colonial Intervention and the Evolution of Water Supply in Benin City, Nigeria." *Water History* 4 (3): 215–229.

Acquaah-Harrison, Richard. 2004. *Housing and Urban Development in Ghana: With Special Reference to Low-Income Housing*. Nairobi: UN-Habitat.

ACTogether. 2015. "Mbale City Profiling Report." Accessed January 12, 2018. https://sdinet.org/wp-content/uploads/2015/04/Mbale_Profile_11.pdf.

Adalet, Begüm. 2022. "Infrastructures of Decolonization: Scales of Worldmaking in the Writings of Frantz Fanon." *Political Theory* 50 (1): 5–31.

Addie, Jean-Paul D. 2022. "The Times of Splintering Urbanism." *Journal of Urban Technology* 29 (1): 109–116.

Adelekan, Ibidun O. 2020. "Urban Dynamics, Everyday Hazards and Disaster Risks in Ibadan, Nigeria." *Environment and Urbanization* 32 (1): 213–232.

Adow, Mohamed. 2020. "The Climate Debt: What the West Owes the Rest." *Foreign Affairs* 99 (3): 60–68.

Adshead, Daniel, Scott Thacker, Lena I. Fuldauer, and Jim W. Hall. 2019. "Delivering on the Sustainable Development Goals through Long-Term Infrastructure Planning." *Global Environmental Change* 59 (November): 101975.

Adunbi, Omolade, and Bilal Butt. 2019. "Afro-Chinese Engagements: Infrastructure, Land, Labour and Finance Introduction." *Africa* 89 (4): 633–637.

AENOR. 2010. "Validation Report of the POA for Uganda Waste to Compost Program." Accessed January 9, 2019. https://cdm.unfccc.int/ProgrammeOfActivities/poa_db /JL4B8R2DKF9ONE6YXCVOQ3MWSGT5UA/V7Q8JRD1BKHUZ9S4L2OMIG5NFP6XYO /ReviewInitialComments/GLDOBAGRHOJT4IVWRZCTN8UV6KSJT4.

Afenah, Afia. 2012. "Engineering a Millennium City in Accra, Ghana: The Old Fadama Intractable Issue." *Urban Forum* 23 (4): 527–540.

Africa Check. 2018. "China Owns 21.3% of Kenya's External Debt—Not 70% as Reported." Accessed October 21, 2019. https://africacheck.org/fact-checks/reports/china -owns-213-kenyas-external-debt-not-70-reported.

African Development Bank. 2011. "The Middle of the Pyramid: Dynamics of the Middle Class in Africa" Accessed January 7, 2020. https://www.afdb.org/sites/default /files/documents/publications/the_middle_of_the_pyramid_the_middle_of_the _pyramid.pdf.

African Development Bank. 2018. "Africa's Infrastructure: Great Potential but Little Impact on Inclusive Growth." Accessed March 6, 2019. https://www.afdb .org/fileadmin/uploads/afdb/Documents/Publications/2018AEO/African_Economic _Outlook_2018_-_EN_Chapter3.pdf.

African Development Bank Group. 2023. "Multinational—LAPSSET—Lamu Port-South Sudan-Ethiopia Project." Accessed June 18, 2023. https://projectsportal.afdb .org/dataportal/VProject/show/P-Z1-DD0-010.

African News Agency. 2016. "Write Off Municipal Debts of Polokwane's Poor: EFF." Accessed July 24, 2019. https://www.iol.co.za/news/politics/write-off-municipal-debts -of-polokwanes-poor-eff-2032237.

African Review of Business and Technology. 2012. "USUSD 100 Million Invested in Ghana City Developments." Accessed July 20, 2019. http://www.africanreview .com/construction-a-mining/buildings/us100-million-investment-in-ghana-city -development.

African Union. 2017. "Infrastructure Corridors Are Key to Africa's Intra-regional Trade, Job Creation: Stakeholders Agree at PIDA Session." Accessed June 18, 2021. https://au.int/sw/node/33483.

African Union. n.d. "Agenda 2063: The Africa We Want." Accessed March 22, 2020. https://au.int/en/agenda2063/overview.

Agergaard, Jytte, Susanne Kirkegaard, and Torben Birch-Thomsen. 2021. "Between Village and Town: Small-Town Urbanism in Sub-Saharan Africa." *Sustainability* 13 (3): 1417.

Ajibade, Idowu. 2017. "Can a Future City Enhance Urban Resilience and Sustainability? A Political Ecology Analysis of Eko Atlantic City, Nigeria." *International Journal of Disaster Risk Reduction* 26: 85–92.

Ajimotokan, Olawale. 2018. "AfCFTA: Dangote Cement Consumption in Africa to Attain 275.7 MT By 2030." *This Day*, July 9, 2018. https://www.thisdaylive.com /index.php/2018/07/09/afcfta-dangote-cement-consumption-in-africa-to-attain-275 -7-mt-by-2030-2/.

Ajl, Max. 2021. "A People's Green New Deal: Obstacles and Prospects." *Agrarian South: Journal of Political Economy* 10 (2): 371–390.

Akyeampong, Emmanuel. 2018. "African Socialism; or the Search for an Indigenous Model." *Economic History of Developing Regions* 33 (1): 69–87.

Alba, Rossella, Antje Bruns, Lara Esther Bartels, and Michelle Kooy. 2019. "Water Brokers: Exploring Urban Water Governance through the Practices of Tanker Water Supply in Accra." *Water* 11 (9): 1919.

Alda-Vidal, Cecilia, Michelle Kooy, and Maria Rusca. 2018. "Mapping Operation and Maintenance: An Everyday Urbanism Analysis of Inequalities within Piped Water Supply in Lilongwe, Malawi." *Urban Geography* 39 (1): 104–121.

Alzouma, Gado. 2005. "Myths of Digital Technology in Africa: Leapfrogging Development?" *Global Media and Communication* 1 (3): 339–356.

Amankwaa, Ebenezer F., and Katherine V. Gough. 2022. "Everyday Contours and Politics of Infrastructure: Informal Governance of Electricity Access in Urban Ghana." *Urban Studies* 59 (12): 2468–2488.

Amankwah-Amoah, Joseph. 2015. "Solar Energy in Sub-Saharan Africa: The Challenges and Opportunities of Technological Leapfrogging." *Thunderbird International Business Review* 57 (1): 15–31.

American Water Works Association. 2012. "Buried No Longer: Confronting America's Water Infrastructure Challenge." Accessed February 16, 2018. http://www.awwa .org/Portals/0/files/legreg/documents/BuriedNoLonger.pdf.

Amin, Samir. 1974. "Accumulation and Development: A Theoretical Model." *Review of African Political Economy* 1 (1): 9–26.

Amin, Samir. 2006. "The Millennium Development Goals: A Critique from the South." *Monthly Review* 57 (10). Accessed 12 January, 2020. https://monthlyreview .org/2006/03/01/the-millennium-development-goals-a-critique-from-the-south/.

Amuzu, David. 2018. "Environmental Injustice of Informal E-waste Recycling in Agbogbloshie-Accra: Urban Political Ecology Perspective." *Local Environment* 23 (6): 603–618.

Anand, Nikhil. 2017. *Hydraulic City: Water and the Infrastructures of Citizenship in Mumbai*. Durham, NC: Duke University Press.

Angelo, Hillary, and David Wachsmuth. 2015. "Urbanizing Urban Political Ecology: A Critique of Methodological Cityism." *International Journal of Urban and Regional Research* 39 (1): 16–27.

Apostolopoulou, Elia. 2021. "Tracing the Links Between Infrastructure-Led Development, Urban Transformation, and Inequality in China's Belt and Road Initiative." *Antipode* 53 (3): 831–858.

Appiah-Kubi, Kojo. 2001. "State-Owned Enterprises and Privatisation in Ghana." *The Journal of Modern African Studies* 39 (2): 197–229.

Appolonia. n.d. "Why Choose Appolonia City." Accessed June 21, 2020. https://www.appolonia.com.gh/why-choose-appolonia/.

Arboleda, Martín. 2020. *Planetary Mine: Territories of Extraction under Late Capitalism*. London: Verso Books.

Arefin, Mohammed Rafi. 2019. "Infrastructural Discontent in the Sanitary City: Waste, Revolt, and Repression in Cairo." *Antipode* 51 (4): 1057–1078.

Asante, Lewis Abedi, and Ilse Helbrecht. 2020. "The Urban Dimension of Chinese Infrastructure Finance in Africa: A Case of the Kotokuraba Market Project, Cape Coast, Ghana." *Journal of Urban Affairs* 42 (8): 1278–1298.

Ashcroft, Bill. 2009. "Alternative Modernities: Globalization and the Post-Colonial." *ARIEL: A Review of International English Literature* 40 (1): 81–105.

Asiedu, Alex Boakye, and Godwin Arku. 2009. "The Rise of Gated Housing Estates in Ghana: Empirical Insights from Three Communities in Metropolitan Accra." *Journal of Housing and the Built Environment* 24 (3): 227–247.

Bah, El-hadj M., Issa Faye, and Zekebweliwai F. Geh. 2018. *Housing Market Dynamics in Africa*. London: Palgrave Macmillan.

Bakker, Karen. 2003. "Archipelagos and Networks: Urbanization and Water Privatization in the South." *Geographical Journal* 169 (4): 328–341.

Baptista, Idalina. 2015. "'We Live on Estimates': Everyday Practices of Prepaid Electricity and the Urban Condition in Maputo, Mozambique." *International Journal of Urban and Regional Research* 39 (5): 1004–1019.

Baptista, Idalina. 2018. "Space and Energy Transitions in Sub-Saharan Africa: Understated Historical Connections." *Energy Research & Social Science* 36: 30–35.

Baptista, Idalina. 2019. "Electricity Services Always in the Making: Informality and the Work of Infrastructure Maintenance and Repair in an African City." *Urban Studies* 56 (3): 510–525.

Baptista, Idalina, and Liza Rose Cirolia. 2022. "From Problematisation to Propositionality: Advancing Southern Urban Infrastructure Debates." *Transactions of the Institute of British Geographers* 47 (4): 927–939.

Barns, Sarah. 2019. "Negotiating the Platform Pivot: From Participatory Digital Ecosystems to Infrastructures of Everyday Life." *Geography Compass* 13 (9): e12464.

Bartels, Lara Esther, Antje Bruns, and Rossella Alba. 2018. "The Production of Uneven Access to Land and Water in Peri-urban Spaces: De Facto Privatisation in Greater Accra." *Local Environment* 23 (12): 1172–1189.

Bayart, Jean-François. 1993. *The State in Africa: The Politics of the Belly*. London: Wiley.

Bayart, Jean-François, and Stephen Ellis. 2000. "Africa in the World: A History of Extraversion." *African Affairs* 99 (395): 217–267.

Bayat, Asef. 2000. "From 'Dangerous Classes' to 'Quiet Rebels': Politics of the Urban Subaltern in the Global South." *International Sociology* 15 (3): 533–557.

Bayırbağ, Mustafa Kemal, Jonathan S. Davies, and Sybille Muench. 2017. "Interrogating Urban Crisis: Cities in the Governance and Contestation of Austerity." *Urban Studies* 54 (9): 2023–2038.

Becker, Sören, and Matthias Naumann. 2017. "Energy Democracy: Mapping the Debate on Energy Alternatives." *Geography Compass* 11 (8): e12321.

Beckert, Sven. 2015. *Empire of Cotton: A Global History*. London: Vintage.

Benjamin, Walter. 1997. *Charles Baudelaire: A Lyric Poet in the Era of High Capitalism*. London: Verso.

Berman, Marshall. 1983. *All That Is Solid Melts into Air: The Experience of Modernity*. London: Verso.

Berrisford, Stephen, Liza Rose Cirolia, and Ian Palmer. 2018. "Land-Based Financing in Sub-Saharan African Cities." *Environment and Urbanization* 30 (1): 35–52.

Bhambra, Gurminder. 2007. *Rethinking Modernity: Postcolonialism and the Sociological Imagination*. London: Springer.

Bicknell-Hersco, Prilly. 2020. "Reparations in the Caribbean and Diaspora." *Caribbean Quilt* 5: 35–43.

Blake, William. 2004. *The Complete Poems*. London: Penguin

Boa, Elizabeth. 1996. *Kafka: Gender, Class, and Race in the Letters and Fictions*. Oxford: Oxford University Press.

Boakye-Ansah, Akosua Sarpong, Klaas Schwartz, and Margreet Zwarteveen. 2019. "From Rowdy Cartels to Organized Ones? The Transfer of Power in Urban Water Supply in Kenya." *The European Journal of Development Research* 31 (5): 1246–1262.

Bond, Patrick. 2010. "Climate Debt Owed to Africa: What to Demand and How to Collect?" *African Journal of Science, Technology, Innovation and Development* 2 (1): 83–113.

Bond, Patrick. 2013. "Water Rights, Commons and Advocacy Narratives." *South African Journal on Human Rights* 29 (1): 125–143.

Bouzarovski, Stefan. 2014. "Energy Poverty in the European Union: Landscapes of Vulnerability." *Wiley Interdisciplinary Reviews: Energy and Environment* 3 (3): 276–289.

Bremner, Lindsay. 2013. "Towards a Minor Global Architecture at Lamu, Kenya." *Social Dynamics* 39 (3): 397–413.

Briant Carant, Jane. 2017. "Unheard Voices: A Critical Discourse Analysis of the Millennium Development Goals' Evolution into the Sustainable Development Goals." *Third World Quarterly* 38 (1): 16–41.

Brown, Julian. 2015. *South Africa's Insurgent Citizens: On Dissent and the Possibility of Politics*. London: Bloomsbury Publishing.

Brown, Stephanie. 2014. "Planning Kampala: Histories of Sanitary Intervention and In/formal Spaces." *Critical African Studies* 6 (1): 71–90.

Brown-Luthango, Mercy. 2021. "An Excluded and Unrecognized Majority: Everyday Struggles of Backyarders in the Western Area of the Voortrekker Road Corridor in Cape Town, South Africa." In *Inclusive Urban Development in the Global South*, edited by Andrea Rigon and Vanesa Castán Broto, 45–60. New York: Routledge.

Bulkeley, Harriet, and Vanesa Castán Broto. 2013. "Government by Experiment? Global Cities and the Governing of Climate Change." *Transactions of the Institute of British Geographers* 38 (3): 361–375.

Burra, Sundar, Sheela Patel, and Thomas Kerr. 2003. "Community-Designed, Built and Managed Toilet Blocks in Indian Cities." *Environment and Urbanization* 15 (2): 11–32.

Camara, Babacar. 2005. "The Falsity of Hegel's Theses on Africa." *Journal of Black Studies* 36 (1): 82–96.

Cardoso, Ricardo. 2016. "The Circuitries of Spectral Urbanism: Looking Underneath Fantasies in Luanda's New Centralities." *Urbanisation* 1 (2): 95–113.

CARE. 2007. *Climate Change and Poverty in Ghana*. Accra: CARE.

Caricom, n.d. "10 Point Reparation Plan." Accessed 12 September 2020. https://caricomreparations.org/caricom/caricoms-10-point-reparation-plan/.

Carmody, Pádraig, and Alicia Fortuin. 2019. "'Ride-Sharing,' Virtual Capital and Impacts on Labor in Cape Town, South Africa." *African Geographical Review* 38 (3): 196–208.

Castán Broto, Vanesa. 2019. *Urban Energy Landscapes*. Cambridge: Cambridge University Press.

Castán Broto, Vanesa, and Harriet Bulkeley. 2013. "Maintaining Climate Change Experiments: Urban Political Ecology and the Everyday Reconfiguration of Urban Infrastructure." *International Journal of Urban and Regional Research* 37 (6): 1934–1948.

Castán Broto, Vanesa, and H. S. Sudhira. 2019. "Engineering Modernity: Water, Electricity and the Infrastructure Landscapes of Bangalore, India." *Urban Studies* 56 (11): 2261–2279.

Centers for Disease Control and Prevention. 2011. Legionellosis—United States, 2000–2009. *Morbidity and Mortality Weekly Report* 60 (32): 1083–1086.

Central Corridor Transit Transport Facilitation Agency. 2018. "About Us." Accessed January 18, 2019. https://centralcorridor-ttfa.org/overview/.

Chakrabarty, Dipesh. 2002. *Habitations of Modernity: Essays in the Wake of Subaltern Studies*. Chicago: University of Chicago Press.

Chakrabarty, Dipesh. 2008. *Provincializing Europe: Postcolonial Thought and Historical Difference*. Princeton: Princeton University Press.

Chalfin, Brenda. 2010. *Neoliberal Frontiers*. Chicago: University of Chicago Press.

Chalfin, Brenda. 2014. "Public Things, Excremental Politics, and the Infrastructure of Bare Life in Ghana's City of Tema." *American Ethnologist* 41 (1): 92–109.

Chambers, Iain. 2017. *Postcolonial Interruptions, Unauthorised Modernities*. London. Rowman & Littlefield.

Chambers, Joseph, and James Evans. 2020. "Informal Urbanism and the Internet of Things: Reliability, Trust and the Reconfiguration of Infrastructure." *Urban Studies* 57 (14): 2918–2935.

Chaplin, Susan E. 1999. "Cities, Sewers and Poverty: India's Politics of Sanitation." *Environment and Urbanization* 11 (1): 145–158.

Charitonidou, Marianna. 2021. "Housing Programs for the Poor in Addis Ababa: Urban Commons as a Bridge between Spatial and Social." *Journal of Urban History* 48 (6): 1345–1364.

Chigudu, Simukai. 2019. "The Politics of Cholera, Crisis and Citizenship in Urban Zimbabwe: 'People Were Dying Like Flies.'" *African Affairs* 118 (472): 413–434.

Chimee, Nkemjika. 2020. "The Transformative Power of European Technology in Resource Exploitation: Reflections on the Oil Presses and Railways of Colonial Nigeria." *Global Environment* 13 (3): 555–582.

Chingono, Mark, and Steve Nakana. 2009. "The Challenges of Regional Integration in Southern Africa." *African Journal of Political Science and International Relations* 3 (10): 396–408.

Cirolia, Liza Rose. 2020. "Fractured Fiscal Authority and Fragmented Infrastructures: Financing Sustainable Urban Development in Sub-Saharan Africa." *Habitat International* 104: 102233.

Cirolia, Liza Rose, and James C. Mizes. 2019. "Property Tax in African Secondary Cities: Insights from the Cases of Kisumu (Kenya) and M'Bour (Senegal)." ICTD Working Paper 90, Institute of Development Studies, Brighton, UK.

Cirolia, Liza Rose, and Suraya Scheba. 2019. "Towards a Multi-scalar Reading of Informality in Delft, South Africa: Weaving the 'Everyday' with Wider Structural Tracings." *Urban Studies* 56 (3): 594–611.

City of Polokwane. 2021. "Annual Budget of Polokwane Municipality 2021/22–2023/24." Accessed January 12, 2022. https://leetolapolokwane.co.za/wp-content/uploads/2021/06/LIM354-MTREF-BUDGET-OF-2021_22-TO-2023_24.pdf.

Climate Focus. 2011. "The Handbook for Programs of Activities." Accessed January 19, 2019. https://climatefocus.com/publications/handbook-programme-activities-2nd-edition/

Coletta, Claudio, and Rob Kitchin. 2017. "Algorhythmic Governance: Regulating the 'Heartbeat' of a City Using the Internet of Things." *Big Data & Society* 4 (2): 1–16 https://doi.org/10.1177/2053951717742418.

Collingswood Water Department. 2017. "Annual Drinking Water Quality Report." Accessed May 12, 2018. http://cms6.revize.com/revize/collingswoodnj/document_center/Water%20Department/Water%20Quality%20Report%202017.pdf.

Conrad, Joseph. 1902. *Heart of Darkness*. London: Penguin.

Cornea, Natasha Lee, René Véron, and Anna Zimmer. 2017. "Everyday Governance and Urban Environments: Towards a More Interdisciplinary Urban Political Ecology." *Geography Compass* 11 (4): e12310.

Cotterrill, Joseph. 2015. "Uber Drivers in South Africa's Gauteng Province Air Unhappiness over Fare Share." *Financial Times*, May 13, 2017. https://www.ft.com/content/dcd36470-206c-11e7-b7d3-163f5a7f229c.

Coutard, Olivier, and Jonathan Rutherford. 2011. "The Rise of Post-networked Cities in Europe? Recombining Infrastructural, Ecological and Urban Transformations in Low Carbon Transitions." In *Cities and Low Carbon Transitions*, edited by Harriet Bulkeley, Vanesa Castán Broto, Mike Hodson, and Simon Marvin, 123–141. London: Routledge.

Coutard, Olivier, and Jonathan Rutherford, eds. 2015. *Beyond the Networked City: Infrastructure Reconfigurations and Urban Change in the North and South*. London: Routledge.

Cowen, Deborah. 2017. "Infrastructures of Empire and Resistance." Accessed October 17, 2020. https://www.versobooks.com/blogs/3067-infrastructures-of-empire-and-resistance.

Cowie, Jefferson. 2001. *Capital Moves: RCA's Seventy-Year Quest for Cheap Labor*. Ithaca, NY: Cornell University Press.

Cramm, Jane M., Xander Koolman, Valerie Møller, and Anna P. Nieboer. 2011. "Socio-economic Status and Self-Reported Tuberculosis: A Multilevel Analysis in a Low-Income Township in the Eastern Cape, South Africa." *Journal of Public Health in Africa* 2 (2): e34.

Cronon, William. 1991. *Nature's Metropolis*. New York: W. W. Norton.

Curtis, Mark and Tim Jones. 2017. "Honest Accounts 2017 How the World Profits From Africa's Wealth." Accessed 15 January 2020. https://debtjustice.org.uk/wp-content /uploads/2017/05/Honest-Accounts-2017-WEB-FINAL.pdf

CSIS. 2018. "Urbanization in Sub-Saharan Africa: Meeting Challenges by Bridging Stakeholders." Accessed January 28, 2018. https://www.csis.org/analysis/urbanization -sub-saharan-africa.

Jean Cuvelier and Laurent de Lucques. 1953. *Relations sur le Congo du Pére 1800-1717*. Brussels, Institut Eoyal Xolonial Belge.

Cumbers, Andrew. 2012. *Reclaiming Public Ownership: Making Space for Economic Democracy*. London: Zed Books.

Cupers, Kenny, and Prita Meier. 2020. "Infrastructure between Statehood and Self-hood: The Trans-African highway." *Journal of the Society of Architectural Historians* 79 (1): 61–81.

Dakyaga, Francis, Sophie Schramm, and Alphounce G. Kyessi. 2022. "Between Self-Help and Emerging Water Markets: Self-Governance, Everyday Practices and the Spatiality of Water Access in Dar es Salaam." *Urban Geography*. https://doi.org/10 .1080/02723638.2022.2106054.

Datta, Ayona. 2016. "Introduction: Fast Cities in an Urban Age." In *Mega-Urbanization in the Global South: Fast Cities and New Urban Utopias of the Postcolonial State*, edited by Ayona Datta and Abdul Shaban, 13–40. Abingdon, UK: Routledge.

Datta, Ayona. 2019. "Postcolonial Urban Futures: Imagining and Governing India's Smart Urban Age." *Environment and Planning D: Society and Space* 37 (3): 393–410.

Datta, Ayona, and Nancy Odendaal. 2011 "Smart cities and the banality of power." *Environment and Planning D: Society and Space* 37 (3): 387–392.

Davies, Jonathan S., and Ismael Blanco. 2017. "Austerity Urbanism: Patterns of Neo-liberalisation and Resistance in Six Cities of Spain and the UK." *Environment and Planning A* 49 (7): 1517–1536.

Dawson, Katherine. 2021. "Geologising Urban Political Ecology (UPE): The Urban-isation of Sand in Accra, Ghana." *Antipode* 53 (4): 995–1017.

De Boeck, Filip. 2011. "Inhabiting Ocular Ground: Kinshasa's Future in the Light of Congo's Spectral Urban Politics." *Cultural Anthropology* 26 (2): 263–286.

De Boeck, Filip. 2015. "'Divining' the City: Rhythm, Amalgamation and Knotting as Forms of 'Ubanity.'" *Social Dynamics* 41 (1): 47–58.

De Boeck, Filip, Ann Cassiman, and Steven Van Wolputte. 2010. "Recentering the City: An Anthropology of Secondary Cities in Africa." Accessed 19 January, 2020. https://repository.up.ac.za/handle/2263/59948.

De Medeiros, Emo. n.d. "Vodunaut." Accessed May 17, 2022. https://www.emodeme deiros.com/vodunaut-1/ko666rkd9hthm1n2peemsy15bl2kad.

Demissie, Alexander. 2018. "Special Economic Zones: Integrating African Countries in China's Belt and Road Initiative." In *Rethinking the Silk Road*, edited by Maximilian Mayer, 69–84. Singapore: Palgrave Macmillan.

Demissie, Fassil. 2007. "Imperial Legacies and Postcolonial Predicaments: An Introduction." *African Identities* 5 (2): 155–165.

Demissie, Fassil, ed. 2012. *Colonial Architecture and Urbanism in Africa: Intertwined and Contested Histories*. London: Ashgate Publishing.

Desai, Renu, Colin McFarlane, and Stephen Graham. 2015. "The Politics of Open Defecation: Informality, Body, and Infrastructure in Mumbai." *Antipode* 47 (1): 98–120.

Dickson, Kwamina B. 1969. *A Historical Geography of Ghana*. Cambridge: Cambridge University Press.

Dienel, Hans-Liudger. 2016. *Linking Networks: The Formation of Common Standards and Visions for Infrastructure Development*. London: Routledge.

Di Nunzio, Marco. 2019. *The Act of Living: Street Life, Marginality, and Development in Urban Ethiopia*. Ithaca, NY: Cornell University Press.

Diop Cheikh Anta. 1974. *The African Origin of Civilization: Myth or Reality*. Chicago: Chicago Review Press.

Doherty, Jacob. 2017. "Life (and Limb) in the Fast-Lane: Disposable People as Infrastructure in Kampala's Boda Boda Industry." *Critical African Studies* 9 (2): 192–209.

Doshi, Sapana. 2017. "Embodied Urban Political Ecology: Five Propositions." *Area* 49 (1): 125–128.

Duhau, Emilio. 1992. "La régularisation de l'habitat au Mexique [The Regulation of Habitat in Mexico]." *Annales de la Recherche Urbaine* 51:48–55.

Dupuy, Gabriel, and Joel Arthur Tarr. 1988. *Technology and the Rise of the Networked City in Europe and America*. Philadelphia: Temple University Press.

Easterling, Keller. 2014. *Extrastatecraft: The Power of Infrastructure Space*. London: Verso Books.

Edwards, Paul N. 2003. "Infrastructure and Modernity: Force, Time, and Social Organization in the History of Sociotechnical Systems." In *Modernity and Technology*, edited by Thomas J. Misa, Philip Brey, and Andrew Feenberg, 185. Cambridge, MA: MIT Press.

Ehwi, Richmond Juvenile, and Nicky Morrison. 2022. "Entanglements in Urban Governance in New African Cities: Appolonia City in the Greater Accra Region, Ghana." *Journal of Urban Affairs* 1–21. doi.org/10.1080/07352166.2022.2074855

Eisenstadt, Shmuel N. 2000. "Multiple Modernities." *Daedalus* 129: 1.

Eisenstadt, Shmuel N., ed. 2017. *Multiple Modernities.* London: Routledge.

Ejigu, Alazar G. 2014. "History, Modernity, and the Making of an African Spatiality: Addis Ababa in Perspective." *Urban Forum* 25 (3): 267–293.

Eko Atlantic. n.d. "About." Accessed July 23, 2020. https://www.ekoatlantic.com /about-us/.

Engels, Frederich. 1892. *The Condition of the Working Class in England in 1844.* London: S. Sonnenschein & Co.

Enns, Charis. 2018. "Mobilizing Research on Africa's Development Corridors." *Geoforum* 88: 105–108.

Enns, Charis, and Brock Bersaglio. 2020. "On the Coloniality of 'New' Mega-Infrastructure Projects in East Africa." *Antipode* 52 (1): 101–123.

Enwezor, Okwui. 2010. "Modernity and Postcolonial Ambivalence." *South Atlantic Quarterly* 109 (3): 595–620.

EPA. 1994. "Combined Sewer Overflow (CSO) Control Policy." Accessed January 12, 2020. https://www.epa.gov/sites/default/files/2015-10/documents/owm0111.pdf.

EPA. 2000. Ghana National Submission to UNFCCC. Accra, Ghana National Government. Accessed March 29, 2020. https://unfccc.int/resource/docs/natc/ghanc3.pdf.

EPA. 2018. A Wet Weather Case Study Incorporating Community Interests in Effective Infrastructure. Accessed 12 January, 2020. https://navigatetheflood.org/resources /a-wet-weather-case-study-of-incorporating-community-interests-into-effective -infrastructure-decision-makings/

Eshun, Maame Esi, and Joe Amoako-Tuffour. 2016. "A Review of the Trends in Ghana's Power Sector." *Energy, Sustainability and Society* 6 (1): 1–9.

Esquivel, Valeria., 2016. Power and the Sustainable Development Goals: a Feminist Analysis. *Gender & Development,* 24 (1): 9–23.

Fält, Lena. 2019. "New Cities and the Emergence of 'Privatized Urbanism' in Ghana." *Built Environment* 44 (4): 438–460.

Fanon, Frantz. 1967. *The Wretched of the Earth.* London: Penguin.

Federici, Silvia. 2019. "Women, Reproduction, and the Commons." *South Atlantic Quarterly* 118 (4): 711–724.

Ferguson, James. 1999. *Expectations of Modernity*. Berkeley: University of California Press.

Ferguson, James. 2006. *Global Shadows: Africa in the Neoliberal World Order*. Durham, NC, Duke University Press.

Fink, Günther, John R. Weeks, and Allan G. Hill. 2012. "Income and Health in Accra, Ghana: Results from a Time Use and Health Study." *The American Journal of Tropical Medicine and Hygiene* 87 (4): 608.

Fitch, Bob, and Mary Oppenheimer. 1966. *Ghana: End of an Illusion*. New York: Monthly Review Press.

Fobil, Julius Najah, and Daniel K. Attuquayefio. 2003. "Remediation of the Environmental Impacts of the Akosombo and Kpong Dams." Accessed 24 August 2021. http://solutions-site.org/node/76.

Food and Water Watch. 2010. "Has water privatization gone too far in New Jersey?" Accessed 12 October 2021. https://www.inthepublicinterest.org/wp-content/uploads/FWW_PrivatizationInNewJersey.pdf.

Franklin, V. P. 2013. "Commentary—Reparations as a Development Strategy: The CARICOM Reparations Commission." *Journal of African American History* 98 (3): 363–366.

Fuchs, Christian, and Eva Horak. 2008. "Africa and the Digital Divide." *Telematics and Informatics* 25 (2): 99–116.

Furlong, Kathryn. 2014. "STS beyond the 'Modern Infrastructure Ideal': Extending Theory by Engaging with Infrastructure Challenges in the South." *Technology in Society* 38: 139–147.

Furlong, Kathryn, and Michelle Kooy. 2017. "Worlding Water Supply: Thinking Beyond the Network in Jakarta." *International Journal of Urban and Regional Research* 41 (6): 888–903.

Gaber, Nadia, Andrew Silva, Monica Lewis-Patrick, Emily Kutil, Debra Taylor, and Roslyn Bouier. 2021. "Water Insecurity and Psychosocial Distress: Case Study of the Detroit Water Shutoffs." *Journal of Public Health* 43 (4): 839–845.

Gandy, Matthew. 2005. "Cyborg Urbanization: Complexity and Monstrosity in the Contemporary City." *International Journal of Urban and Regional Research* 29 (1): 26–49.

Gandy, Matthew. 2014. *The Fabric of Space: Water, Modernity, and the Urban Imagination*. Cambridge, MA: MIT Press.

Ghana Statistical Service. 2012. *Population and Housing Census: Summary Report of Final Results*. Accra: Sakoa Press.

Ghana Statistical Service. 2022. Accessed March 3, 2022. https://census2021.statsghana.gov.gh/

Gidwani, Vinay, and Amita Baviskar. 2011. "Urban Commons." *Economic and Political Weekly* 46 (50): 42–43.

Gillespie, Tom. 2016. "Accumulation by Urban Dispossession: Struggles over Urban Space in Accra, Ghana." *Transactions of the Institute of British Geographers* 41 (1): 66–77.

Gillespie, Tom. 2020. "The Real Estate Frontier." *International Journal of Urban and Regional Research* 44 (4): 599–616.

Gillette, Howard, Jr. 2005. *Camden after the Fall*. Philadelphia: University of Pennsylvania Press.

Gilroy, Paul. 1993. *The Black Atlantic: Modernity and Double Consciousness*. Cambridge, MA: Harvard University Press.

Githaiga, Nancy Muthoni, and Wang Bing. 2019. "Belt and Road Initiative in Africa: The Impact of Standard Gauge Railway in Kenya." *China Report* 55 (3): 219–240.

GIZ. n.d. "Improving the Infrastructure in Africa." Accessed September 20, 2021. https://www.giz.de/en/worldwide/28079.html.

GIZ. 2019. "Access to Water and Sanitation in Sub-Saharan Africa." Accessed 12 April 2021. https://www.oecd.org/water/GIZ_2018_Access_Study_Part%20I_Synthesis_Report.pdf.

Gollin, Doug. 2018. *Structural Transformation and Growth Without Industrialisation*. Oxford: Pathways for Prosperity Commission.

Goodell, Jeff. 2018. *The Water Will Come: Rising Seas, Sinking Cities and the Remaking of the Civilized World*. Melbourne, Australia: Black Inc.

Goodfellow, Tom. 2017. "Urban Fortunes and Skeleton Cityscapes: Real Estate and Late Urbanization in Kigali and Addis Ababa." *International Journal of Urban and Regional Research* 41 (5): 786–803.

Goodfellow, Tom. 2020. "Finance, Infrastructure and Urban Capital: The Political Economy of African 'Gap-Filling.'" *Review of African Political Economy* 47 (164): 256–274.

Gotham, Kevin Fox. 2006. "The Secondary Circuit of Capital Reconsidered: Globalization and the US Real Estate Sector." *American Journal of Sociology* 112 (1): 231–275.

Goulding, Richard, Adam Leaver, and Jonathan Silver. 2023. "From Homes to Assets: Transcalar Territorial Networks and the Financialisation of Build to Rent in Greater Manchester." *Environment and Planning A. 0308518X221138104*.

Government of Kenya. n.d. "Vision 2030." Accessed 12 October 2020. https://vision2030.go.ke/

Graham, Stephen. 2009. "Cities as Battlespace: The New Military Urbanism." *City* 13 (4): 383–402.

Graham, Stephen, and Colin McFarlane, eds. 2014. *Infrastructural Lives*. London: Taylor & Francis.

Graham, Stephen, and Simon Marvin. 2001. *Splintering Urbanism: Networked Infrastructures, Technological Mobilities and the Urban Condition*. London: Routledge.

Grant, Richard. 2009. *Globalizing City: The Urban and Economic Transformation of Accra, Ghana*. Syracuse, NY: Syracuse University Press.

Green, Toby. 2019. *A Fistful of Shells: West Africa from the Rise of the Slave Trade to the Age of Revolution*. London: Penguin.

Guma, Prince K. 2019. "Smart Urbanism? ICTs for Water and Electricity Supply in Nairobi." *Urban Studies* 56 (11): 2333–2352.

Guma, Prince K. 2020. "Incompleteness of Urban Infrastructures in Transition: Scenarios from the Mobile Age in Nairobi." *Social Studies of Science* 50 (5): 728–750.

Guma, Prince.K. and Monstadt, Jochen. 2021. "Smart city making? The spread of ICT-driven plans and infrastructures in Nairobi". *Urban Geography*, 42(3): 360–381.

Guma, Prince K., Jochen Monstadt, and Sophie Schramm. 2019. "Hybrid Constellations of Water Access in the Digital Age: The Case of Jisomee Mita in Soweto-Kayole, Nairobi." *Water Alternatives* 12 (2): 725–743.

Guma, Prince K., Jochen Monstadt, and Sophie Schramm. 2022. "Post-, Pre- and Non-payment: Conflicting Rationalities in the Digitalisation of Energy Access in Kibera, Nairobi." *Digital Geography and Society* 3: 100037.

Gwaindepi, Abel. 2019. "Serving God and Mammon: The 'Minerals-Railway Complex' and Its Effects on Colonial Public Finances in the British Cape Colony, 1810–1910." African Economic History Working Paper Series 44.

Gyau-Boakye, Phillip. 2001. "Environmental Impacts of the Akosombo Dam and Effects of Climate Change on the Lake Levels." *Environment, Development and Sustainability* 3: 17–29.

Gyau-Boakye, Frank. 2021. "The Mortgage Landscape in Ghana." Accessed April 2 2022. http://landeconomy.knust.edu.gh/news/news-articles/mortgage-landscape-ghana.

Hakam, Sabrine Salima. 2019. "Examining 'Cityness' to Inform an Urban Theory of the 'Global South.'" PhD diss., King's College London, UK.

Hall, David, Emanuele Lobina, and Violeta Corral. 2010. *Replacing Failed Private Water Contracts*. Ferney-Voltaire, France: Public Services International (PSI).

Hamer, John H. 1981. "Preconditions and Limits in the Formation of Associations: The Self-Help and Cooperative Movement in Sub-Saharan Africa." *African Studies Review* 24 (1): 113–132.

Hansard. 1900. "Uganda Railway Bill." Accessed March 3, 2022. https://hansard.parliament.uk/Lords/1900-06-21/debates/402c8a08-f46f-41e7-bc2f-36263a5c38be/UgandaRailwayBill.

Hardoy, Jorge E., and David Satterthwaite, eds. 2019. *Small and Intermediate Urban Centers*. London: Routledge.

Harvey, David. 1978. "The Urban Process Under Capitalism: A Framework for Analysis." *International Journal of Urban and Regional Research* 2 (1–3): 101–131.

Harvey, David. 1996. *Justice, Nature and the Politics of Difference*. New York: Wiley-Blackwell.

Harvey, David. 2001. "Globalization and the 'Spatial Fix.'" *Geographische Revue: Zeitschrift für Literatur und Diskussion* 3 (2): 23–30.

Harvey, David. 2007. *A Brief History of Neoliberalism*. Oxford: Oxford University Press.

Hegel, Georg Wilhelm Friedrich. [1837] 1956. *The Philosophy of History*. New York: Dover Publications.

Hess, Janet Berry. 2000. "Imagining Architecture: The Structure of Nationalism in Accra, Ghana." *Africa Today* 47 (2): 35–58.

Heynen, Nik, Maria Kaika, and Erik Swyngedouw, eds. 2006. *In the Nature of Cities: Urban Political Ecology and the Politics of Urban Metabolism*. London: Routledge.

Hodos, Jerome. 2011. *Second Cities: Globalization and Local Politics in Manchester and Philadelphia*. Philadelphia: Temple University Press.

Hodson, Mike, and Simon Marvin. 2009. "'Urban Ecological Security': A New Urban Paradigm?" *International Journal of Urban and Regional Research* 33 (1): 193–215.

Holston, James. 2009. "Insurgent Citizenship in an Era of Global Urban Peripheries." *City & Society* 21 (2): 245–267.

Hoornweg, Daniel, and Kevin Pope. 2017. "Population Predictions for the World's Largest Cities in the 21st Century." *Environment and Urbanization* 29 (1): 195–216.

Howard-Hassmann, Rhoda E. 2018. *Reparations to Africa*. Philadelphia: University of Pennsylvania Press.

Hughes, Thomas. 1983. *Networks of Power: Electrification in Western Society, 1880–1930*. Baltimore: John Hopkins University Press.

Hyman, Katherine, and Edgar Pieterse. 2017. "Infrastructure Deficits and Potential in African Cities." In *The SAGE Handbook of the 21st Century City*, edited by Suzanne Hall and Ricky Burdett, 429–451. London: Sage.

Ibrahim, Basil, and Amiel Bize. 2018. "Waiting Together: The Motorcycle Taxi Stand as Nairobi Infrastructure." *Africa Today* 65 (2): 73–92.

IEA. 2021. "Access to Electricity." Accessed 12 April 2022. https://www.iea.org/reports/sdg7-data-and-projections/access-to-electricity.

Iliffe, John. 1987. *The African Poor: A History*. Cambridge: Cambridge University Press.

International Labor Organization. "2015. SWTS Country Brief: Uganda." Accessed July 7, 2020 https://www.ilo.org/wcmsp5/groups/public/@ed_emp/documents/publica tion/wcms_429078~1.pdf

Index for Censorship. 2016. "Joint Letter on Internet Shutdown in Uganda." Accessed July 5, 2019. https://www.indexoncensorship.org/2016/02/joint-letter-on -internet-shutdown-in-uganda/.

Izar, Priscila. 2022. "Meanings of Self-Building: Incrementality, Emplacement, and Erasure in Dar es Salaam's Traditional Swahili Neighborhoods." *Urban Planning* 7 (1): 305–320.

Jabary Salamanca, Omar, and Jonathan Silver. 2022. "In the Excess of Splintering Urbanism: The Racialized Political Economy of Infrastructure." *Journal of Urban Technology* 29 (1): 117–125.

Jaglin, Sylvy. 2008. "Differentiating Networked Services in Cape Town: Echoes of Splintering Urbanism?" *Geoforum* 39 (6): 1897–1906.

Jaglin, Sylvy. 2014. "Rethinking Urban Heterogeneity." In *The Routledge Handbook on Cities of the Global South*, edited by Susan Parnell and Sophie Oldfield, 434–446. London: Routledge.

Jaglin, Sylvy. 2015. "Is the Network Challenged by the Pragmatic Turn in African Cities." In *Beyond the Networked City: Infrastructure Reconfigurations and Urban Change in the North and South*, edited by Olivier Coutard and Jonathan Rutherford, 182–203. London: Routledge.

Jaglin, Sylvy. 2019. "Electricity Autonomy and Power Grids in Africa: From Rural Experiments to Urban Hybridizations." *Local Energy Autonomy: Spaces, Scales, Politics* 1: 291–314.

Jambadu, Lazarus, Alfred Dongzagla, and Ishmael Kabange. 2022. "Understanding Intra-urban Inequality in Networked Water Supply in Wa, Ghana." *GeoJournal*. https://doi.org/10.1007/s10708-022-10662-z.

Jambadu, Lazarus, Jochen Monstadt, and Sophie Schramm. 2022. "Understanding Repair and Maintenance in Networked Water Supply in Accra and Dar es Salaam." *Water Alternatives* 15 (2): 265–289.

James, Cyril Lionel Robert. 1938. *The Black Jacobins*. New York: Vintage.

Janu, Swati. 2017. "On-the-Go Settlements: Understanding Urban Informality through Its Digital Substructure." *Urbanisation* 2 (2): 116–134.

Jarosz, Lucy. 1992. "Constructing the Dark Continent: Metaphor as Geographic Representation of Africa." *Geografiska Annaler: Series B, Human Geography* 74 (2): 105–115.

Jedwab, Remi, Edward Kerby, and Alexander Moradi. 2017. "History, Path Dependence and Development: Evidence from Colonial Railways, Settlers and Cities in Kenya." *The Economic Journal* 127 (60): 1467–1494.

Kafka, Franz. 1926. *The Castle*. London: Penguin.

Kaika, Maria. 2005. *City of Flows: Modernity, Nature, and the City*. London: Routledge.

Kaika, Maria, and Erik Swyngedouw. 2000. "Fetishizing the Modern City: The Phantasmagoria of Urban Technological Networks." *International Journal of Urban and Regional Research* 24 (1): 120–138.

Kamalu, Chukwunyere. 2018. *The Word at Face Value—An Abridged Account of Dogon Cosmology*. London: Orisa Press.

Kanai, J. Miguel, Richard Grant, and Radu Jianu. 2018. "Cities on and off the Map: A Bibliometric Assessment of Urban Globalisation Research." *Urban Studies* 55 (12): 2569–2585.

Kandji, Serigne Tacko, Louis Verchot, and Jens Mackensen. 2006. "Climate Change and Variability in the Sahel Region: Impacts and Adaptation Strategies in the Agricultural sector." Accessed October 15, 2021. http://apps.worldagroforestry.org /downloads/Publications/PDFS/B14549.pdf.

Karuri-Sebina, Geci, Karel-Herman Haegeman, and Apiwat Ratanawaraha. 2016. "Urban Futures: Anticipating a World of Cities." *Foresight* 18 (5): 449–453.

Kasanga, R. Kasim, and Nii Ashie Kotey. 2001. *Land Management in Ghana: Building on Tradition and Modernity*. London: IIED.

Katz, Cindi. 2001. "Vagabond Capitalism and the Necessity of Social Reproduction." *Antipode* 33 (4): 709–728.

Kazeem, A. 2020. "African Mobile Users Pay Nearly Three Times the Global Average for Voice Calls and Internet." Accessed October 13, 2021. https://qz.com/africa /1878749/how-much-do-africans-pay-for-voice-calls/.

Kazibwe, Kenneth. N.d. "Uber Drivers Strike over Pay, Mistreatment by Company." Accessed September 22, 2019. http://nilepost.co.ug/2019/09/24/uber-drivers-strike -over-pay-mistreatment-by-company/.

Kilaka, Benard Musembi, and Jan Bachmann. 2021. "Kenya Launches Lamu Port. But Its Value Remains an Open Question." Accessed January 14, 2022. https://theconversation .com/kenya-launches-lamu-port-but-its-value-remains-an-open-question-161301.

Killingray, David. 1986. "The Maintenance of Law and Order in British Colonial Africa." *African Affairs* 85 (340): 411–347.

Kimari, Wangui. 2020. "War-Talk: An Urban Youth Language of Siege in Nairobi." *Journal of Eastern African Studies* 14 (4): 707–723.

Kimari, Wangui. 2021. "The Story of a Pump: Life, Death and Afterlives within an Urban Planning of 'Divide and Rule' in Nairobi, Kenya." *Urban Geography* 42 (2): 141–160.

Kimari, Wangui, and Henrik Ernstson. 2020. "Imperial Remains and Imperial Invitations: Centering Race Within the Contemporary Large-Scale Infrastructures of East Africa." *Antipode* 52 (3): 825–846.

King, Anthony D. 1977. "Exporting 'Planning': The Colonial and Neo-Colonial Experience." *Urbanism Past & Present* 5:12–22.

Kitchin, Rob, and Martin Dodge. 2019. "The (In)Security of Smart Cities: Vulnerabilities, Risks, Mitigation, and Prevention." *Journal of Urban Technology* 26 (2): 47–65.

Klaufus, Christien. 2010. "Watching the City Grow: Remittances and Sprawl in Intermediate Central American Cities." *Environment and Urbanization* 22 (1): 125–137.

Klein, Naomi. 2007. *The Shock Doctrine: The Rise of Disaster Capitalism.* New York: Picador.

Knight Frank. 2015. "Africa Report 2015: Real Estate Markets in a Continent of Growth and Prosperity." Accessed November 13, 2021. https://content.knightfrank .com/research/155/documents/en/africa-report-2015-2802.pdf.

Konadu-Agyemang, Kwadwo. 2001. *The Political Economy of Housing and Urban Development in Africa: Ghana's Experience from Colonial Times to 1998.* London: Praeger.

Kooy, Michelle. 2014. "Developing Informality: The Production of Jakarta's Urban Waterscape." *Water Alternatives* 7 (1): 35–53.

Kooy, Michelle, and Karen Bakker. 2008. "Technologies of Government: Constituting Subjectivities, Spaces, and Infrastructures in Colonial and Contemporary Jakarta." *International Journal of Urban and Regional Research* 32 (2): 375–391.

Kuykendall, Ronald. 1993. "Hegel and Africa: An Evaluation of the Treatment of Africa in the Philosophy of History." *Journal of Black Studies* 23 (4): 571–581.

Lang, M. 2017. *Trans-Saharan Trade Routes.* New York: Cavendish Square.

Lawanson, Taibat, Taofiki Salau, and Omoayena Yadua. 2013. "Conceptualizing the Liveable African City." *Journal of Construction Project Management and Innovation* 3 (1): 573–588.

Lawhon, Mary. 2013. "Flows, Friction and the Sociomaterial Metabolization of Alcohol." *Antipode* 45 (3): 681–701.

Lawhon, Mary, David Nilsson, Jonathan Silver, Henrik Ernstson, and Shuaib Lwasa. 2018. "Thinking Through Heterogeneous Infrastructure Configurations." *Urban Studies* 55 (4): 720–732.

Lawhon, Mary, Henrik Ernstson, and Jonathan Silver. 2014. "Provincializing Urban Political Ecology: Towards a Situated UPE through African Urbanism." *Antipode* 46 (2): 497–516.

Lemanski, Charlotte. 2020. "Infrastructural Citizenship: The Everyday Citizenships of Adapting and/or Destroying Public Infrastructure in Cape Town, South Africa." *Transactions of the Institute of British Geographers* 45 (3): 589–605.

Lesutis, Gediminas. 2020. "How to Understand a Development Corridor? The Case of Lamu Port–South Sudan–Ethiopia–Transport Corridor in Kenya." *Area* 52 (3): 600–608.

Levontin, Polina. 2020. "African Contemporary Artists and SF." Accessed January 13, 2021. https://vector-bsfa.com/2020/11/24/african-contemporary-artists-and-sf/.

Loftus, Alex. 2012. *Everyday Environmentalism: Creating an Urban Political Ecology*. Minneapolis: University of Minnesota Press.

Lokanathan, Venkateswaran. 2020. "China's Belt and Road Initiative: Implications in Africa." Observer Research Foundation Brief 395.

Lutter, Mark. 2019. "The 3 Trends Shaping the Future of Africa's Cities." Accessed September 13, 2021. https://www.theafricareport.com/20886/the-3-trends-shaping-the-future-of-africas-cities/.

Lwasa, Shuaib. 2010. "Adapting Urban Areas in Africa to Climate Change: The Case of Kampala." *Current Opinion in Environmental Sustainability* 2 (3): 166–171.

Lwasa, Shuaib. 2014. "Managing African Urbanization in the Context of Environmental Change." *Interdisciplina* 2 (2). https://doi.org/10.22201/ceiich.24485705e.2014.2.46528.

Mabin, Alan, Siân Butcher, and Robin Bloch. 2013. "Peripheries, suburbanisms and change in sub-Saharan African cities." *Social Dynamics* 39 (2): 167–190.

Mabunda, Tiyani E., Nalezani J. Ramalivhana, and Yoswa M. Dambisya. 2012. "Mortality associated with tuberculosis/HIV co-infection among patients on TB treatment in the Limpopo province, South Africa." *African Health Sciences* 14 (4): 849–854.

Macamo, Elsio Salvado, ed. 2005. *Negotiating Modernity: Africa's Ambivalent Experience*. London: Zed Books.

Macey, David. 2012. *Frantz Fanon: A Biography*. London: Verso Books.

Mack, Elizabeth A., and Sarah Wrase. 2017. "A Burgeoning Crisis? A Nationwide Assessment of the Geography of Water Affordability in the United States." *PLoS One* 12 (1): e0169488.

Mackinder, Halford J. 1900. "A Journey to the Summit of Mount Kenya, British East Africa." *The Geographical Journal* 15 (5): 453–476.

Mahajan, Vijay. 2011. *Africa Rising: How 900 Million African Consumers Offer More Than You Think*. Hoboken, NJ: Pearson Prentice Hall.

MainOne. n.d. "MainOne Expands Data Center Footprint in Ghana." Accessed July 12, 2020. https://www.mainone.net/ghana/blog/2019/10/03/mainone-expands-data-center-footprint-in-ghana-3/.

Marais, Lochner, and Jan Cloete. 2017. "The Role of Secondary Cities in Managing Urbanisation in South Africa." *Development Southern Africa* 34 (2): 182–195.

Marcuse, Peter. 1997. "The Enclave, the Citadel, and the Ghetto: What Has Changed in the Post-Fordist US City." *Urban Affairs Review* 33 (2): 228–264.

Marvin, Simon, and Jonathan Rutherford. 2018. "Controlled Environments: An Urban Research Agenda on Microclimatic Enclosure." *Urban Studies* 55 (6): 1143–1162.

Marx, Karl. [1867] 1990. *Capital: A Critique of Political Economy*. Vol. 1. London: Penguin.

Matagi, Samuel Vivian. 2002. "Some Issues of Environmental Concern in Kampala, the Capital City of Uganda." *Environmental Monitoring and Assessment* 77 (2): 121–138.

Mathis, Joel. 2005. "Is Camden Really America's Most Dangerous City?" Accessed May 12, 2019. https://www.phillymag.com/news/2015/02/04/camden-really-americas -dangerous-city/.

Mattern Shannon. 2018. "Maintenance and Care" Accessed 22 September 2020 https://placesjournal.org/article/maintenance-and-care/.

Maw, Peter, Terry Wyke, and Alan Kidd. 2012. "Canals, Rivers, and the Industrial City: Manchester's Industrial Waterfront, 1790–1850." *The Economic History Review* 65 (4): 1495–1523.

Mazrui, Ali. 1999. "From Slave Ship to Space Ship: Africa between Marginalization and Globalization." *African Studies Quarterly* 2 (4): 5–11.

Mazzucato, Mariana. 2011. "The Entrepreneurial State." *Soundings* 49: 131–142.

Mbale Municipality. 2010. *Mbale Development Plan*. Mbale: Mbale Municipality.

Mbembe, Achille. 1992. "The Banality of Power and the Aesthetics of Vulgarity in the Postcolony." *Public Culture* 4 (2): 1–30.

Mbembe, Achille. 2017. *Critique of Black Reason*. Durham, NC: Duke University Press.

Mbembe, Achille. 2018. "The Idea of a Borderless World." Accessed April 15, 2021. https://chimurengachronic.co.za/the-idea-of-a-borderless-world/.

Mbembe, Achille, and Sarah Nuttall. 2004. "Writing the World from an African Metropolis." *Public Culture* 16 (3): 347–372.

Mboya, Tom. 1963. "African Socialism." *Transition* 8: 17–19.

McDonald, David. 2009. *Electric Capitalism: Recolonising Africa on the Power Grid*. London: Routledge.

McFarlane, Colin. 2008. "Governing the Contaminated City: Infrastructure and Sanitation in Colonial and Post-Colonial Bombay." *International Journal of Urban and Regional Research* 32 (2): 415–435.

McFarlane, Colin, and Jonathan Silver. 2017. "Navigating the City: Dialectics of Everyday Urbanism." *Transactions of the Institute of British Geographers* 42 (3): 458–471.

Medici. 2019. "M-Pesa: Mobile Phone-Based Money Transfer—Global Presence." Accessed November 19, 2020. https://gomedici.com/m-pesa-mobile-phone-based -money-transfer-global-presence.

Meldrum, Andrew. 2004. "Mobile Phones the Talk of Africa as Landlines Lose Out." *The Guardian*, May 5, 2004. Accessed September 29, 2020. https://www.theguardian.com/media/2004/may/05/citynews.newmedia.

Melosi, Martin V. 2000. *Effluent America: Cities, Industry, Energy, and the Environment.* Pittsburgh: University of Pittsburgh Press.

Mendelsohn, Ben. 2018. "Making the Urban Coast: A Geosocial Reading of Land, Sand, and Water in Lagos, Nigeria." *Comparative Studies of South Asia, Africa and the Middle East* 38 (3): 455–472.

Mercer, Claire. 2014. "Middle Class Construction: Domestic Architecture, Aesthetics and Anxieties in Tanzania." *The Journal of Modern African Studies* 52 (2): 227–250.

Mercer, Claire. 2017. "Landscapes of Extended Ruralisation: Postcolonial Suburbs in Dar es Salaam, Tanzania." *Transactions of the Institute of British Geographers* 42 (1): 72–83.

Merrifield, Andy. 2002. *Dialectical Urbanism.* New York: Monthly Review Press.

Meth, Paula, Tom Goodfellow, Alison Todes, and Sarah Charlton. 2021. "Conceptualizing African Urban Peripheries." *International Journal of Urban and Regional Research* 45 (6): 985–1007.

Mhlongo, Niq. 2011. *After Tears.* Cape Town: Kwela Books.

Millington, Nate, and Suraya Scheba. 2021. "Day Zero and the Infrastructures of Climate Change: Water Governance, Inequality, and Infrastructural Politics in Cape Town's Water Crisis." *International Journal of Urban and Regional Research* 45 (1): 116–132.

Minuchin, Leandro. 2021. "Prefigurative Urbanization: Politics Through Infrastructural Repertoires in Guayaquil." *Political Geography* 85: 102316.

Misa, Thomas J., Philip Brey, and Andrew Feenberg, eds. 2003. *Modernity and Technology.* Cambridge, MA: MIT Press.

Mitchell, Timothy, ed. 2000. *Questions of Modernity.* Minneapolis: University of Minnesota Press.

Mkutu, Kennedy, Marie Müller-Koné, and Evelyne Atieno Owino. 2021. "Future Visions, Present Conflicts: The Ethnicized Politics of Anticipation Surrounding an Infrastructure Corridor in Northern Kenya." *Journal of Eastern African Studies* 15 (4): 707–727.

Mohan, Jitendra. 1966. "Varieties of African Socialism." *Socialist Register* 3. https://socialistregister.com/index.php/srv/article/view/5974.

Monson, Jamie. 2009. *Africa's Freedom Railway: How a Chinese Development Project Changed Lives and Livelihoods in Tanzania.* Bloomington: Indiana University Press.

Monstadt, Jochen, and Sophie Schramm. 2017. "Toward the Networked City? Translating Technological Ideals and Planning Models in Water and Sanitation Systems in Dar es Salaam." *International Journal of Urban and Regional Research* 41 (1): 104–125.

Monteith, William. 2019. "Markets and Monarchs: Indigenous Urbanism in Postcolonial Kampala." *Settler Colonial Studies* 9 (2): 247–265.

Moore, Jason. 2015. *Capitalism in the Web of Life: Ecology and the Accumulation of Capital*. London: Verso Books.

Morange, Marianne, Fabrice Folio, Elisabeth Peyroux, and Jeanne Vivet. 2012. "The Spread of a Transnational Model: 'Gated Communities' in Three Southern African Cities (Cape Town, Maputo and Windhoek)." *International Journal of Urban and Regional Research* 36 (5): 890–914.

Moss, Timothy. 2020. *Remaking Berlin: A History of the City Through Infrastructure, 1920–2020*. Cambridge, MA: MIT Press.

Muddu, Iai. 2015. "A City of Two Tales." In *Boda Boda Anthem and Other Poems: A Kampala Poetry Anthology*, edited by Mildred Kiconco Barya, 68. Kampala: Babishai Niwe Poetry Foundation.

Mudimbe, Valentin-Yves. 1988. *The Invention of Africa: Gnosis, Philosophy, and the Order of Knowledge*. Bloomington: Indiana University Press.

Muhumuza, William. 2008. "Between Rhetoric and Political Conviction: The Dynamics of Decentralization in Uganda and Africa." *Journal of Social, Political, and Economic Studies* 33 (4): 426–457.

Mulenga, Gadzeni. 2013. "'Developing Economic Corridors' in Africa Rationale for the Participation of the African Development Bank." Accessed September 12, 2020. https://www.afdb.org/fileadmin/uploads/afdb/Documents/Publications/Regional_Integration_Brief_-_Developing_Economic_Corridors_in_Africa_-_Rationale_for_the_Participation_of_the_AfDB.pdf.

Munro, Paul. 2020. "On, off, below and beyond the Urban Electrical Grid the Energy Bricoleurs of Gulu Town." *Urban Geography* 41 (3): 428–447.

Musemwa, Muchaparara. 2010. "From 'Sunshine City' to a Landscape of Disaster: The Politics of Water, Sanitation and Disease in Harare, Zimbabwe, 1980–2009." *Journal of Developing Societies* 26 (2): 165–206.

Muwanga, David. 2018. "Port Bell to Kampala rail line due to re-open." *The Observer*, April 23, 2018 Accessed September 19, 2021.

Meldrum, Andrew. 2004. "Mobile Phones the Talk of Africa as Landlines Lose Out." *The Guardian*, May 5, 2004. Accessed September 29, 2020.

Mugisha, Joshua, Mike Arasa Ratemo, Bienvenu Christian Bunani Keza, and Hayriye Kahveci. 2021. "Assessing the opportunities and challenges facing the development of off-grid solar systems in Eastern Africa: The cases of Kenya, Ethiopia, and Rwanda." *Energy Policy* 150: 112131.

Myambo, Melissa Tandiwe. 2014. "Imagining a Dialectical African Modernity: Achebe's Ontological Hopes, Sembene's Machines, Mda's Epistemological Redness." *Journal of Contemporary African Studies* 32 (4): 457–473.

Myers, Garth. 2006. "The Unauthorized City: Late Colonial Lusaka and Postcolonial Geography." *Singapore Journal of Tropical Geography* 27 (3): 289–308.

Myers, Garth. 2008. "Peri-urban Land Reform, Political-Economic Reform, and Urban Political Ecology in Zanzibar." *Urban Geography* 29 (3): 264–288.

Myers, Garth. 2011. *African Cities: Alternative Visions of Urban Theory and Practice.* London: Zed Books.

Myers, Garth. 2014. "From Expected to Unexpected Comparisons: Changing the Flows of Ideas about Cities in a Postcolonial Urban World." *Singapore Journal of Tropical Geography* 35 (1): 104–118.

Myers, Garth. 2018. "The Africa problem of global urban theory: re-conceptualising planetary urbanisation". In *African Cities and the Development Conundrum,* edited by Carole Ammann and Till Förster, 231–253. Leiden, Brill.

Myers, Garth. 2020. *Rethinking Urbanism: Lessons from Postcolonialism and the Global South.* Bristol: Bristol University Press.

Naisanga, Priscilla. n.d. "Drop 0.5% on Mobile Money (MM)Tax." Accessed October 12, 2019. https://www.newvision.co.ug/news/1483174/drop-05-mobile-money-tax.

Nakaweesi, David. 2017. "East Africa: Kampala-Dar es Salaam—Another Route Uganda Should Consider?" *The Monitor,* March 8, 2017. Accessed September 15, 2019. https://allafrica.com/stories/201703080199.html.

Nakyagaba, Gloria Nsangi, Mary Lawhon, Shuaib Lwasa, Jonathan Silver, and Fredrick Tumwine. 2021. "Power, Politics and a Poo Pump: Contestation over Legitimacy, Access and Benefits of Sanitation Technology in Kampala." *Singapore Journal of Tropical Geography* 42 (3): 415–430.

Narsiah, Sagie. 2011. "The Struggle for Water, Life, and Dignity in South African Cities: The Case of Johannesburg." *Urban Geography* 32 (2): 149–155.

Nation. 2016. "Yoweri Museveni Explains Social Media, Mobile Money Shutdown." Accessed April 6, 2021. https://nation.africa/kenya/news/Yoweri-Museveni-explains-social-media-mobile-money-shutdown/1056-3083032-8h5ykhz/index.html.

Nation. 2020. "East Africa Now Owes China USD 29.4 Billion in Infrastructure Loans." Accessed January 12, 2021. https://nation.africa/kenya/news/africa/East-Africa-now-owes-China-29-billion-in-infrastructure-loans/1066-4807662-bspmmkz/index.html.

National Planning Authority. 2013. "Uganda Vision 2040." Accessed April 26, 2021. http://www.npa.go.ug/uganda-vision-2040/.

Ncube, Mthuli., and Charles Lufumpa. 2014. *The Emerging Middle Class in Africa.* London: Routledge.

Ndlovu-Gatsheni, J., and Brilliant Mhlanga, eds. 2013. *Bondage of Boundaries and Identity Politics in Postcolonial Africa: The Northern Problem and Ethno-Futures.* Oxford: African Books Collective.

Ndung'u, Njuguna. 2018. "The M-Pesa Technological Revolution for Financial Services in Kenya: A Platform for Financial Inclusion." In *Handbook of Blockchain, Digital Finance, and Inclusion*, edited by David Lee Kuo Chuen and Robert H. Deng, 37–56. Cambridge, MA: Academic Press.

NEMA. 2012. *The Abbreviated Resettlement Plans (ARAPs) for the Municipal Solid Waste Composting Project in Mbale Municipality.* Kampala: NEMA.

Nemser, Daniel. 2017. *Infrastructures of Race: Concentration and Biopolitics in Colonial Mexico.* Austin: University of Texas Press.

Neves Alves, Susana. 2021. "Everyday States and Water Infrastructure: Insights from a Small Secondary City in Africa, Bafatá in Guinea-Bissau." *Environment and Planning C: Politics and Space* 39 (2): 247–264.

New African. 2018. "The Volta River Project: How Kwame Nkrumah's Dream Project Was Frustrated." Accessed 19 July 2020. https://newafricanmagazine.com/16321/.

Newhouse, Leonie, and Abdou Maliq Simone. 2017. "An Introduction: Inhabiting the Corridor: Surging Resource Economies and Urban Life in East Africa." Accessed 19 January, 2018. https://pure.mpg.de/rest/items/item_2486952_1/component/file_2486951/content.

New Jersey Office of the State Comptroller. 2009. "State Comptroller finds Camden's mismanagement of water contract cost taxpayers millions." Accessed 12 May 2018. https://www.state.nj.us/comptroller/news/ docs/amden_pr_final12-16-09.pdf/.

New Jersey Futures. 2014. "Camden Rising above the Flood Waters." Accessed May 12, 2018. https://www.njfuture.org/wp-content/uploads/2014/05/Camden__Final.pdf.

Ngwenya, Nobukhosi, and Liza Rose Cirolia. 2021. "Conflicts Between and Within: The 'Conflicting Rationalities' of Informal Occupation in South Africa." *Planning Theory & Practice* 22 (5): 691–706.

Nicholls, Christine Stephanie. 1971. *The Swahili Coast: Politics, Diplomacy and Trade on the East African Littoral, 1798–1856.* New York: Africana Publishing Corporation.

Nilsson, David. 2016. "The Unseeing State: How Ideals of Modernity Have Undermined Innovation in Africa's Urban Water Systems." *NTM Zeitschrift für Geschichte der Wissenschaften, Technik und Medizin* 24 (4): 481–510.

Njeru, Jeremia. 2006. "The Urban Political Ecology of Plastic Bag Waste Problem in Nairobi, Kenya." *Geoforum* 37 (6): 1046–1058.

Njoh, Ambe J. 2008. "Colonial Philosophies, Urban Space, and Racial Segregation in British and French Colonial Africa." *Journal of Black Studies* 38 (4): 579–599.

Nkosi, Ntombi. 2021. "Water shortages plague Polokwane, residents blame government." *IOL*, October 4, 2021. Accessed November 21, 2022.

Nkrumah, Kwame. 1965. *Neo-colonialism: The Last Stage of Imperialism*. London: Nelson.

Nkrumah, Kwame. 1997. *Selected Speeches*. Vol. 1. Accra: Afram Publications Ghana.

Nyamai, Dorcas Nthoki, and Sophie Schramm. 2022. "Accessibility, Mobility, and Spatial Justice in Nairobi, Kenya." *Journal of Urban Affairs* 1–23. https://doi.org/10.1080/07352166.2022.2071284.

Nyamnjoh, Francis B. 2001. "Expectations of Modernity in Africa or a Future in the Rear-View Mirror?" *Journal of Southern African Studies* 27 (2): 363–369.

Nye, David. 1992. *Electrifying America: Social Meanings of a New Technology*. Cambridge, MA: MIT Press.

Nyerere, Julius. 1964. "Freedom and Unity." *Transition* 14: 40–45.

Nziza, Richard, Mbaga-Niwampa, and David Mukholi. 2011. *Peoples and Cultures of Uganda*. Kampala: Fountain House.

O'Connor, Anthony Michael. 1965. *Railways and Development in Uganda: A Study in Economic Geography*. Oxford: Oxford University Press.

Obeng-Odoom,, Franklin. 2013a. "Africa's Failed Economic Development Trajectory: A Critique." *African Review of Economics and Finance* 4 (2): 151–175.

Obeng-Odoom, Franklin. 2013b. "The State of African Cities 2010: Governance, Inequality and Urban Land Markets." *Cities* 31 (2): 425–429.

Obeng-Odoom, Franklin. 2014. "Green Neoliberalism: Recycling and Sustainable Urban Development in Sekondi-Takoradi." *Habitat International* 41: 129–134.

Obeng-Odoom, Franklin. 2015a. "Africa: On the Rise, but to Where?" *Forum for Social Economics* 44 (3): 234–250.

Obeng-Odoom, Franklin. 2015b. "Informal Real Estate Brokerage as a Socially-Embedded Market for Economic Development in Africa." In *Real Estate, Construction and Economic Development in Emerging Market Economies*, edited by Raymond Talinbe Abdulai, Franklin Obeng-Odoom, Edward Ochieng, and Vida Maliene. London: Taylor & Francis. 224–239.

Odendaal, Nancy. 2011. "Splintering Urbanism or Split Agendas? Examining the spatial Distribution of Technology Access in Relation to ICT Policy in Durban, South Africa." *Urban Studies* 48(11): 2375–2397.

Odendaal, Nancy. 2016. "Smart City: Neoliberal Discourse or Urban Development Tool?" In *The Palgrave Handbook of International Development*, edited by Jean Grugel and Daniel Hammett, 615–633. London: Palgrave Macmillan.

Odendaal, Nancy. 2018. "Smart Innovation at the Margins: Learning from Cape Town and Kibera." In *Inside Smart Cities*, edited by Andrew Karvonen, Federico Cugurullo, and Federico Caprotti, 243–257. London: Routledge.

Oxford English Dictionary. Accessed December 22, 2020. https://www.oxfordlear nersdictionaries.com/.

Ohemeng, Frank L. K., and Joshual J. Zaato. 2021. "The Failure to Learn Lessons from Policy Failures in Developing Countries? The Case of Electricity Privatization in Ghana." *International Journal of Public Administration* 1–13. https://doi.org/10 .1080/01900692.2021.2001012.

Oji, H. 2019. Dangote Cement Posts N390.3 Billion Full Year Profit. *The Guardian Nigeria*, February 28, 2019. Accessed December 19, 2019. https://guardian.ng /business-services/dangote-cement-posts-n390-3-billion-full-year-profit/.

Okot-Okumu, James, and Richard Nyenje. 2011. "Municipal Solid Waste Management under Decentralisation in Uganda." *Habitat International* 35 (4): 537–543.

Olajide, Oluwafemi, and Taibat Lawanson. 2022. "Urban Paradox and the Rise of the Neoliberal City: Case Study of Lagos, Nigeria." *Urban Studies* 59 (9): 1763–1781.

Olvera, Lourdes Diaz, Didier Plat, and Pascal Pochet. 2013. "The Puzzle of Mobility and Access to the City in Sub-Saharan Africa." *Journal of Transport Geography* 32: 56–64.

Ong, Ai-hwa. 2011. "Introduction: Worlding Cities, or the Art of Being Global." In *Worlding Cities: Asian Experiments and the Art of Being Global*, edited by Ananya Roy and Aihwa Ong. London: John Wiley & Sons.–1–27

Onyishi, Chinedu Josephine, Adaeze UP Ejike-Alieji, Chukwuedozie Kelechukwu Ajaero, Casmir Chukwuka Mbaegbu, Christian Chukwuebuka Ezeibe, Victor Udemezue Onyebueke, Peter Oluchukwu Mbah, and Thaddeus Chidi Nzeadibe. 2021. "COVID-19 Pandemic and Informal Urban Governance in Africa: A Political Economy Perspective." *Journal of Asian and African Studies* 56 (6): 1226–1250.

Oqubay, Arkebe, and Justin Yifu Lin, eds. 2019. *China-Africa and an Economic Transformation*. Oxford: Oxford University Press.

Osabu-Kle, Daniel Tetteh. 2000. "The African Reparation Cry: Rationale, Estimate, Prospects, and Strategies." *Journal of Black Studies* 30 (3): 331–350.

Osborne, George. 2015. "Chancellor Speech in Chengdu, China, on Building a Northern Powerhouse." Accessed August 17, 2020. https://www.gov.uk/government /speeches/chancellor-speech-in-chengdu-china-on-building-a-northern-powerhouse.

Otiso, Kefa M. 2005. "Kenya's Secondary Cities Growth Strategy at a Crossroads: Which Way Forward?" *GeoJournal* 62 (1): 117–128.

Otiso, Kefa M., and George Owusu. 2008. "Comparative Urbanization in Ghana and Kenya in Time and Space." *GeoJournal* 71 (2): 143–157.

Owusu, George. 2018. "The Changing Views on the Role of Small Towns in Rural and Regional Development in Africa." In *Cities in the World, 1500–2000*, edited by Adrian Green, 49–60. London: Routledge.

Parnell, Susan, and Ruwani Walawege. 2011. "Sub-Saharan African Urbanisation and Global Environmental Change." *Global Environmental Change* 21: S12–S20.

Pasaogullari, Hulya. 2019. "Principal Lenders to Uganda." Accessed April 23, 2020. https://assets.publishing.service.gov.uk/media/5d9b3eb9e5274a5a272064c9/591_IFI _External_Lending_Uganda_Kenya.pdf.

Paulais, Thierry. 2012. *Financing Africa's Cities: The Imperative of Local Investment.* DC: World Bank Publications.

Pauw, Jacques. 2003. "The Politics of Underdevelopment: Metered to Death—How a Water Experiment Caused Riots and a Cholera Epidemic." *International Journal of Health Services* 33 (4): 819–830.

Payer, Cheryl. 1975. *The Debt Trap: The International Monetary Fund and the Third World.* New York: New York University Press.

Peck, Jamie. 2012. "Austerity Urbanism: American Cities under Extreme Economy." *City* 16 (6): 626–655.

Pell, M. B., and Joshua Schneyer. 2016. "Thousands of US Areas Afflicted with Lead Poisoning beyond Flint's." Accessed April 13, 2020. https://www.scientificamerican .com/article/thousands-of-u-s-areas-afflicted-with-lead-poisoning-beyond-flints/.

Phillips, Claudette, Jonathan Silver, and Priscilla Rowswell. 2011. "Report on the Mamre Ceiling Insulation Evaluation Energy Retrofitting in Low Income Communities." Accessed 13 January, 2020. https://www.cityenergy.org.za/uploads/resource_354.pdf.

Pieterse, Edgar. 2010. "Cityness and African Urban Development." *Urban Forum* 21 (3): 205–219.

Pieterse, Edgar. 2013. *City Futures: Confronting the Crisis of Urban Development.* London: Bloomsbury Publishing.

Pieterse, Edgar, and Susan Parnell. 2014. *Africa's Urban Revolution.* London: Zed Books.

Pieterse, Edgar, Susan Parnell, and Gareth Haysom. 2018. "African Dreams: Locating Urban Infrastructure in the 2030 Sustainable Developmental Agenda." *Area Development and Policy* 3 (2): 149–169.

Pieterse, Edgar, Susan Parnell, David Simon, and AbdouMaliq Simone. 2010. *Urbanization Imperatives for Africa: Transcending Policy Inertia.* Cape Town: African Center for Cities, University of Cape Town.

Pithouse, Richard. 2014. "An Urban Commons? Notes from South Africa." *Community Development Journal* 49 (1): 31–43.

Ponder, Sage., and Mikael Omstedt. 2022. "The Violence of Municipal Debt: From Interest Rate Swaps to Racialized Harm in the Detroit Water Crisis." *Geoforum* 132: 271–280.

Pow, Choon-Piew. 2011. "Living It Up: Super-Rich Enclave and Transnational Elite Urbanism in Singapore." *Geoforum* 42 (3): 382–393.

Power, Marcus, Giles Mohan, and May Tan-Mullins. 2012. *China's Resource Diplomacy in Africa: Powering Development?* London: Palgrave Macmillan.

Press, Will. 2015. "Monsoons over the Moon." Accessed 12 May, 2020. http://www.bottomline.co.ke/monsoons-over-the-moon-film-review/.

PricewaterhouseCoopers. N.d. "Disrupting Africa: Riding the Wave of the Digital Revolution." Accessed July 18, 2020. https://www.pwc.com/gx/en/issues/high-growth-markets/assets/disrupting-africa-riding-the-wave-of-the-digital-revolution.pdf.

Prowell, George Reeser. 1886. *The History of Camden County, New Jersey*. Philadelphia: Richards.

Public Ledger Philadelphia, Pennsylvania. (1846, January 6). Accessed 12 May, 2020 https://www.newspapers.com/clip/7352462/public_ledger/.

Public Ledger Philadelphia, Pennsylvania. (1853, December 12). Accessed 12 May, 2020 https://www.newspapers.com/clip/7352486/public_ledger/.

Pulido, Laura. 2016. "Flint, Environmental Racism, and Racial Capitalism." *Capitalism Nature Socialism* 27 (3): 1–16.

Quarcoopome, Samuel S. 1993. "A History of the Urban Development of Accra: 1877–1957." *Research Review* 9 (1–2): 20–32.

Ramakrishnan, Kavita, Kathleen O'Reilly, and Jessica Budds. 2021. "The Temporal Fragility of Infrastructure: Theorizing Decay, Maintenance, and Repair." *Environment and Planning E: Nature and Space* 4 (3): 674–695.

Ranganathan, Malini. 2016. "Thinking with Flint: Racial Liberalism and the Roots of an American Water Tragedy." *Capitalism Nature Socialism* 27 (3): 17–33.

Ranganathan, Malini, and Carolina Balazs. 2015. "Water Marginalization at the Urban Fringe: Environmental Justice and Urban Political Ecology Across the North–South Divide." *Urban Geography* 36 (3): 403–423.

Rao, Vyjayanthi. 2014. "Infra-city: Speculations on Flux and History in Infrastructure-Making." In *Infrastructural Lives*, edited by Stephen Graham and Colin McFarlane, 53–72. London: Routledge.

Rendeavour. n.d. "Appolonia City Launches Solar Strategy with Axcon Energy." Accessed July 27, 2020. https://www.rendeavour.com/news/appolonia-city-launches-solar-strategy-with-axcon-energy-solar-farm/.

Resource Center for Energy Economics and Regulation. 2005. "Guide to Electric Power in Ghana." Accessed June 2, 2018. https://pdf.usaid.gov/pdf_docs/Pnads932.pdf.

Reuss, Sophia. 2015. "Understanding the Greek Debt Crisis: Lessons from Structural Adjustment." Accessed August 3, 2015. https://www.alterinter.org/?Understanding-the-Greek-Debt-Crisis-Lessons-from-Structural-Adjustment.

Richardson, Lizzie. 2015. "Performing the Sharing Economy." *Geoforum* 67: 121–129.

Robins, Jonathan. 2016. *Cotton and Race across the Atlantic: Britain, Africa, and America, 1900–1920.* Woodbridge, UK: Boydell & Brewer.

Robins, Steven. 2014. "Poo Wars as Matter out of Place: 'Toilets for Africa' in Cape Town." *Anthropology Today* 30 (1): 1–3.

Robins, Steven. 2019. "'Day Zero,' Hydraulic Citizenship and the Defence of the Commons in Cape Town: A Case Study of the Politics of Water and Its Infrastructures (2017–2018)." *Journal of Southern African Studies* 45 (1): 5–29.

Robinson, Cedric J. 1983. *Black Marxism: The Making of the Black Radical Tradition.* Chapel Hill: University of North Carolina Press.

Robinson, Jennifer. 2002. "Global and World Cities: A View from off the Map." *International Journal of Urban and Regional Research* 26 (3): 531–554.

Robinson, Jennifer. 2006. *Ordinary Cities: Between Modernity and Development.* London: Routledge.

Robinson, Jennifer. 2016. "Thinking Cities through Elsewhere: Comparative Tactics for a More Global Urban Studies." *Progress in Human Geography* 40 (1): 3–29.

Rodney, Walter. 1972. *How Europe Underdeveloped Africa.* London: Verso Books.

Rondinelli, Dennis. 1983. "Towns and Small Cities in Developing Countries." *Geographical Review* 73 (4): 379–395.

Roy, Ananya. 2009. "The 21st-Century Metropolis: New Geographies of Theory." *Regional Studies* 43 (6): 819–830.

Roy, Ananya, and Aihwa Ong, eds. 2011. *Worlding Cities: Asian Experiments and the Art of Being Global.* London: John Wiley & Sons.

Rusca, Maria, and Klaas Schwartz. 2018. "The Paradox of Cost Recovery in Heterogeneous Municipal Water Supply Systems: Ensuring Inclusiveness or Exacerbating Inequalities?" *Habitat International* 73: 101–108.

Sachs, Jeffrey D. 2012. "From Millennium Development Goals to Sustainable Development Goals." *The Lancet* 379 (9832): 2206–2211.

Said, Edward. 1979. *Orientalism.* London: Vintage.

Saito, Fumihiko. 2012. *Decentralization and Development Partnership: Lessons from Uganda.* New York: Springer.

Sambu, Zeddy. 2008. "East Africa: Countries Move to Upgrade Railway Network." Accessed August 13, 2021. http://www.ocnus.net/artman2/publish/Business_1/East _Africa_Countries_Move_to_Upgrade_Railway_Network.shtml.

SAMSET. 2016. "Sustainable Energy in Urban Africa." Accessed September 23, 2021. https://samsetproject.net/wp-content/uploads/2016/02/Sustainable-Energy-in -Urban-Africa-%E2%80%93-the-role-of-local-government-Africities-Summit-2015 -Background-Paper.pdf.

Sankara, Thomas. 1988. *Thomas Sankara Speaks: The Burkina Faso Revolution, 1983-87*. New York: Pathfinder Press.

Sassen, Saskia. 2014. *Expulsions: Brutality and Complexity in the Global Economy*. Cambridge, MA: Harvard University Press.

Satterthwaite, David. 2016. "Background Paper: Small and Intermediate Urban Centers in Sub-Saharan Africa." Urban Africa Risk Knowledge (Urban ARK) Working Paper 6.

Satterthwaite, David. 2017. "Will Africa Have Most of the World's Largest Cities in 2100?" *Environment and Urbanization* 29 (1): 217–220.

Sawyer, L. 2019. "Natures Remade and Imagined: Dreaming 'World City' and Popular Real-Estate in Lagos." In *Grounding Urban Natures: Histories and Futures of Urban Ecologies*, edited by Henrik Ernstson and Svreker Sorlin. Cambridge, MA: MIT Press. 83–107.

Scheba, Suraya. 2021. "Pathways Toward a Dialectical Urbanism: Thinking with the Contingencies of Crisis, Care and Capitalism." In *Global Urbanism*, edited by Michele Lancione and Colin McFarlane, 176–182. London: Routledge.

Schindler, Seth, Jessica DiCarlo, and Dinesh Paudel. 2022. "The New Cold War and the Rise of the 21st-Century Infrastructure State." *Transactions of the Institute of British Geographers* 47 (2): 331–346.

Schindler, Seth, and J. Miguel Kanai. 2021. "Getting the Territory Right: Infrastructure-Led Development and the Re-emergence of Spatial Planning Strategies." *Regional Studies* 55 (1): 40–51.

Schramm, Sophie, and Basil Ibrahim. 2021. "Hacking the Pipes: Hydro-Political Currents in a Nairobi Housing Estate." *Environment and Planning C: Politics and Space* 39 (2): 354–370.

Schramm, Sophie, and Lucía Wright-Contreras. 2017. "Beyond Passive Consumption: Dis/ordering Water Supply and Sanitation at Hanoi's Urban Edge." *Geoforum* 85: 299–310.

Sembene, Ousmane. 1995. *God's Bits of Wood*. London: Heinemann.

Serumaga, Kalundi. 2015. "Greece Birthed European Civilisation but Can No Longer Afford to Live by Its Standards. Who Can Then? Not Africa, Surely." Accessed September 16, 2019. https://ugandansatheart.blogspot.com/2015/06/uah-greece-birthed-european.html.

Shilliam, Robbie. 2016. "Colonial Architecture or Relatable Hinterlands? Locke, Nandy, Fanon and the Bandung Spirit." *Constellations: An International Journal of Critical and Democratic Theory* 23 (3): 425–435.

Shimeles, Abebe, and Mthuli Ncube. 2015. "The Making of the Middle-Class in Africa: Evidence from DHS Data." *The Journal of Development Studies* 51 (2): 178–193.

Silver, Jonathan, and Omar Jabary Salamanca. 2023. "The Freedom Railway: Inscriptions of an Infrastructure Corridor." *New Geographies Journal* 13.

Silver, Jonathan, Cheryl McEwan, Laura Petrella, and Hamidou Baguian. 2013. "Climate Change, Urban Vulnerability and Development in Saint-Louis and Bobo-Dioulasso: Learning from across Two West African Cities." *Local Environment* 18 (6): 663–677.

Simone, AbdouMaliq. 2004a. *For the City Yet to Come: Changing African Life in Four Cities*. Durham, NC: Duke University Press.

Simone, AbdouMaliq. 2004b. "People as Infrastructure: Intersecting Fragments in Johannesburg." *Public Culture* 16 (3): 407–429.

Simone, AbdouMaliq. 2010. *City Life from Jakarta to Dakar: Movements at the Crossroads*. London: Routledge.

Simone, AbdouMaliq. 2013. "Cities of Uncertainty: Jakarta, the Urban Majority, and Inventive Political Technologies." *Theory, Culture & Society* 30 (7–8): 243–263.

Simone, AbdouMaliq. 2020. "Cities of the Global South." *Annual Review of Sociology* 46: 603–622.

Simone, AbdouMaliq. 2022. "A Glossary." Accessed February 14, 2022. https://cityscapesmagazine.com/projects-articles/let-us-begin.

Simone, AbdouMaliq, and Edgar Pieterse. 2018. *New Urban Worlds: Inhabiting Dissonant Times*. London: John Wiley & Sons.

Sinha, Ravi. 2012. "Mutant Modernities." Accessed 12.06.2019. https://countercurrents.org/Mutant_Modernities_Socialist_Futures.pdf.

Slavova, Mira, and Ekene Okwechime. 2016. "African Smart Cities Strategies for Agenda 2063." *Africa Journal of Management* 2 (2): 210–229.

Smiley, Sarah. 2020. "Heterogeneous Water Provision in Dar es Salaam: The Role of Networked Infrastructures and Alternative Systems in Informal Areas." *Environment and Planning E: Nature and Space* 3 (4): 1215–1231.

Smith, Laila. 2001. "The Urban Political Ecology of Water in Cape Town." *Urban Forum* 12 (2): 204–224.

Sokona, Youba, Yacob Mulugetta, and Haruna Gujba. 2012. "Widening Energy Access in Africa: Towards Energy Transition." *Energy Policy* 47: 3–10.

South African Department of Provisional and Local Government. 1998. The White Paper on Local Government, South African Government, Pretoria.

Social Justice Coalition. 2014. "Our Toilets are Dirty: Report of the Social Audit into the Janitorial Service for Communal Flush Toilets in Khayelitsha." Accessed 12 September 2021. https://sjc.org.za/wp-content/uploads/2018/09/social_audit_3_janitorial_service_social_audit_report_final.pdf.

Soyinka, Wole. 1965. *The Interpreters*. London: Penguin.

Spivak, Gayatri. 1993. "Can the Subaltern Speak?" In *Colonial Discourse and Post-Colonial Theory: A Reader*, edited by Patrick. Williams and Laura. Chrisman, 66–112. London: Harvester Wheatsheaf.

Sseviiri, Hakimu, Shuaib Lwasa, Mary Lawhon, Henrik Ernstson, and Revocatus Twinomuhangi. 2022. "Claiming Value in a Heterogeneous Solid Waste Configuration in Kampala." *Urban Geography* 43 (1): 59–80.

Star, Susan Leigh. 1999. "The Ethnography of Infrastructure." *American Behavioural Scientist* 43 (3): 377–391.

State of New Jersey. 1871. "Acts of the Ninety-Firth Legislature." Accessed September 12, 2019. http://www.waterworkshistory.us/NJ/Camden/1871NJchap139.pdf.

Statistics South Africa. 2018. "Non-financial Census of Municipalities, 2017." Accessed September 12, 2019. http://www.statssa.gov.za/?p=11199.

Stoler, Ann Laura. 2016. *Duress: Imperial Durabilities in Our Times*. Durham, NC: Duke University Press.

Streule, Monika, Ozan Karaman, Lindsay Sawyer, and Christian Schmid. 2020. "Popular Urbanization: Conceptualizing Urbanization Processes beyond Informality." *International Journal of Urban and Regional Research* 44 (4): 652–672.

Sundaram, Ravi. 2009. *Pirate Modernity: Delhi's Media Urbanism*. London: Routledge.

Sustainable Energy Africa. 2020. "State of Energy in South African Cities 2020." Accessed October 12, 2021. https://www.sustainable.org.za/uploads/resources/resource_84.pdf.

Swanson, Maynard. 1977. "The Sanitation Syndrome: Bubonic Plague and Urban Native Policy in the Cape Colony, 1900–19091." *The Journal of African History* 18 (3): 387–410.

Swilling, Mark. 2011. "Reconceptualising Urbanism, Ecology and Networked Infrastructures." *Social Dynamics* 37 (1): 78–95.

Swyngedouw, Erik. 2004. *Social Power and the Urbanization of Water: Flows of Power*. Oxford: Oxford University Press.

Swyngedouw, Erik, and Nick Heynen. 2003. "Urban Political Ecology, Justice and the Politics of Scale." *Antipode* 35 (5): 898–918.

Taiwo, Olufemi. 1998. "Exorcising Hegel's Ghost: Africa's Challenge to Philosophy." *African Studies Quarterly* 1 (4): 3–16.

Taiwo, Olufemi. 2010. *How Colonialism Preempted Modernity in Africa*. Bloomington: Indiana University Press.

Tarr, Joel. 1984. "The Evolution of the Urban Infrastructure in the Nineteenth and Twentieth Centuries." In *Perspectives on Urban Infrastructure*, edited by Royce Hanson, 4–66. Washington, DC: National Academies Press.

Tchuwa, Isaac. 2018. The 'Poisoned Chalice'of State Ownership of Water Infrastructure in Contemporary Blantyre, Malawi. *Journal of Contemporary African Studies*, 36 (1): 1–22.

Thieme, Tatiana. 2013. "The 'Hustle' amongst Youth Entrepreneurs in Mathare's Informal Waste Economy." *Journal of Eastern African Studies* 7 (3): 389–412.

Thiong'o, Ngũgĩ Wa. 1977. *Petals of Blood*. London: Heinemann.

Thompson, Tade. 2019. *The Rosewater Insurrection*. London: Orbit.

Tonkiss, Fran. 2013. "Austerity Urbanism and the Makeshift City." *City* 17 (3): 312–324.

Toussaint, Eric. 1999. *Your Money or Your Life: The Tyranny of Global Finance*. London: Pluto Books.

Town and Country Planning Department. 2011. *Structure Plan for the Kasoa-Winneba Corridor: Technical Report*. Accra: Ministry of Local Government, Rural Development and Environment.

Truelove, Yaffa. 2011. "(Re-)Conceptualizing Water Inequality in Delhi, India Through a Feminist Political Ecology Framework." *Geoforum* 42 (2): 143–152.

Truelove, Yaffa. 2019. "Rethinking Water Insecurity, Inequality and Infrastructure through an Embodied Urban Political Ecology." *Water* 6 (3): e1342.

Tuana, Nancy. 2019. "Climate Apartheid: The Forgetting of Race in the Anthropocene." *Critical Philosophy of Race* 7 (1): 1–31.

Twala, Chitja. 2014. "The Causes and Socio-Political Impact of the Service Delivery Protests to the South African Citizenry: A Real Public Discourse." *Journal of Social Sciences* 39 (2): 159–167.

Tzaninis, Yannis, Tait Mandler, Maria Kaika, and Roger Keil. 2021. "Moving Urban Political Ecology beyond the 'Urbanization of Nature.'" *Progress in Human Geography* 45 (2): 229–252.

UN-DESA. 2018. "2018 Revision of World Urbanization Prospects." Accessed September 2, 2019. https://www.un.org/development/desa/publications/2018-revision -of-world-urbanization-prospects.html#:~:text=By%202050%2C%20it%20is%20 projected,to%204.2%20billion%20in%202018.

UN-DESA. 2019. "Polokwane, South Africa Metro Area Population 1950–2022." Accessed April 12, 2020. https://www.macrotrends.net/cities/22498/polokwane/population.

UNFCCC. 2013. "Monitoring Report Form for CDM Program of Activities." Accessed January 14, 2019. https://cdm.unfccc.int/ProgrammeOfActivities/poa_db/JL4B8R2DKF 90NE6YXCVOQ3MWSGT5UA/view.

United Nations General Assembly. 2010. "The human right to water and sanitation milestones." Accessed April 29, 2022. http://www.un.org/waterforlifedecade/pdf /human_right_to_water_and_sanitation_milestones.pdf.

UN-Habitat. 2014. *State of African Cities*. Nairobi: UN-Habitat.

UN-Habitat. 2015. *The Challenge of Local Government Financing in Developing Countries*. Nairobi: UN-Habitat.

UN-Habitat. 2020. "Global Indicators Database." Accessed March 5, 2020. https://data.unhabitat.org/.

UNICEF/UN-Habitat. 2020. "Analysis of Multiple Deprivations in Secondary Cities in Sub-Saharan Africa." Accessed August 15, 2020. https://www.unicef.org/esa/media/5561/file/Analysis%20of%20Multiple%20Deprivations%20in%20Secondary%20Cities%20-%20Analysis%20Report.pdf.

United Nations. 2020. "World Migration Report 2020." Accessed September 14, 2021. https://www.un.org/development/desa/pd/sites/www.un.org.development.desa.pd/files/undesa_pd_2020_international_migration_highlights.pdf.

US Census Bureau. n.d. Accessed October 22, 2020. https://www.census.gov/.

Uwayezu, Ernest, and Walter T. De Vries. 2018. "Indicators for Measuring Spatial Justice and Land Tenure Security for Poor and Low Income Urban Dwellers." *Land* 7 (3): 84.

Van Noorloos, Femke, and Maggi Leung. 2016. "Circulating Asian Urbanisms: An Analysis of Policy and Media Discourse in Africa and Latin America." In *Reconfiguration of the Global South: Africa and Latin America and the "Asian Century"*, edited by Eckart Woertz, 141–158. London: Routledge.

Van Noorloos, Femke, and Marjan Kloosterboer. 2018. "Africa's New Cities: The Contested Future of Urbanisation." *Urban Studies* 55 (6): 1223–1241.

Viljoen, B. 2021. "Polokwane Muni: 'Pay or Get Cut Off.'" Accessed September 22, 2020. https://reviewonline.co.za/536129/no-pay-no-services-2/.

von Schnitzler, Antina. 2013. "Traveling Technologies: Infrastructure, Ethical Regimes, and the Materiality of Politics in South Africa." *Cultural Anthropology* 28 (4): 670–693.

Wahby, Noura. 2021. "Urban Informality and the State: Repairing Cairo's Waters through Gehood Zateya." *Environment and Planning E: Nature and Space* 4 (3): 696–717.

Wambede, Fred, and Phoebe Masongole. 2021. "Mbale Just a City on Paper, Residents Say." *The Monitor*. February 9, 2021. https://www.monitor.co.ug/uganda/news/national/mbale-just-a-city-on-paper-residents-say-3285402.

Wamuchiru, Elizabeth. 2017. "Beyond the Networked City: Situated Practices of Citizenship and Grassroots Agency in Water Infrastructure Provision in the Chamazi settlement, Dar es Salaam." *Environment and Urbanization* 29 (2): 551–566.

Watson, Vanessa. 2014. "African Urban Fantasies: Dreams or Nightmares?" *Environment and Urbanization* 26 (1): 215–231.

Were, Edmond. 2019. "East African Infrastructural Development Race: A Sign of Postmodern Pan-Africanism?" *Cambridge Review of International Affairs* 35 (4): 566–591.

Western Cape Government. 2016. "Enumeration Report: Barcelona Informal Settlement Pocket." Accessed February 22, 2020. https://www.westerncape.gov.za/sites/www.westerncape.gov.za/files/hs-enumeration-study-wcg-ep-barcelona-final.pdf.

Whitman, Walt. 1867. "I Dream'd in a Dream." Accessed January 17, 2020. https://whitmanarchive.org/published/LG/1867/poems/52.

Williams, Patrick. 1999. *Ngugi wa Thiong'o.* Manchester: Manchester University Press.

World Bank. 2010. *A City-Wide Approach to Carbon Finance. Carbon Partnership Facility Innovations Series.* Washington, DC: World Bank.

World Bank. 2022. "International Debt Statistics." Accessed February 12, 2022. https://datatopics.worldbank.org/debt/ids/region/SSA.

World Bank. n.d. "People Using Safely Managed Drinking Water Services, Urban (% of Urban Population)—Sub-Saharan Africa." Accessed February 12, 2020. https://data.worldbank.org/indicator/SH.H2O.SMDW.UR.ZS?locations=ZG.

World Bank. n.d. "Access to electricity, urban (% of urban population) - Sub-Saharan Africa." Accessed January 1, 2022. https://data.worldbank.org/indicator/EG.ELC.ACCS.UR.ZS?locations=ZG&most_recent_year_desc=false.

World Bank. n.d. "What A Waste Global Database." Accessed January 1, 2022. https://datacatalog.worldbank.org/search/dataset/0039597.

World Economic Forum. 2019. "Why Africa's Economic Future Lies in Its Smaller Cities." Accessed 12.04.2021. https://www.weforum.org/agenda/2019/05/putting-africa-s-secondary-cities-first/.

Yeboah, Ian. 2000. "Structural Adjustment and Emerging Urban Form in Accra, Ghana." *Africa Today* 47 (2): 61–89.

Yeoh, Brenda S. A. 2001. "Postcolonial Cities." *Progress in Human Geography* 25 (3): 456–468.

Youé, Chris. 2015. "Berlin 1885: The Division of Africa." *Canadian Journal of African Studies/Revue Canadienne des études Africaines* 49 (2): 439–440.

Zeiderman, Austin, Sobia Ahmad Kaker, Jonathan Silver, and Astrid Wood. 2015. "Uncertainty and Urban Life." *Public Culture* 27 (2): 281–304.

ZiroMwatela, Raphael, and Zhao Changfeng. 2016. "Africa in China's 'One Belt, One Road' Initiative: A Critical Analysis." *IOSR Journal of Humanities and Social Science* 21 (12): 10–21.

SOURCES

Chapter 4 partly draws on work that was originally published as "Incremental infrastructures: Material improvisation and social collaboration across post-colonial Accra" in *Urban Geography* 35, no. 6 (2014): 788–804.

Chapter 5 partly draws on work originally published as "Disrupted infrastructures: An urban political ecology of interrupted electricity in Accra" in the *International Journal of Urban and Regional Research* 39.5 (2015): 984–1003. It also partly drew on research with Colin McFarlane originally published as "The poolitical city: 'Seeing sanitation' and making the urban political in Cape Town." Antipode 49, no. 1 (2017): 125–148.

Chapter 6 also partly draws on work with Colin McFarlane in the paper "Navigating the city: Dialectics of everyday urbanism" in *Transactions of the Institute of British Geographers* 42, no. 3 (2017): 458–471. It also drew on work with Alan Wiig in the paper "Turbulent presents, precarious futures: Urbanization and the deployment of global infrastructure" in *Regional Studies* 53, no. 6 (2019): 912–923.

Chapter 7 partly draws on Jonathan Silver, "The climate crisis, carbon capital and urbanisation: An urban political ecology of low-carbon restructuring in Mbale" in *Environment and Planning A* 49, no. 7 (2017): 1477–1499.

Chapter 9 partly draws on Jonathan Silver, "Decaying infrastructures in the post-industrial city: An urban political ecology of the US pipeline crisis" in *Environment and Planning E: Nature and Space* 4, no. 3 (2019): 756–777.

INDEX

Infrastructures Series

Edited by Paul N. Edwards and Janet Vertesi